Early Modern Literature in History

General Editor: **Cedric C. Brown**
Professor of English and Head of Department, University of Reading

Within the period 1520–1740 this series discusses many kinds of writing, both within and outside the established canon. The volumes may employ different theoretical perspectives, but they share an historical awareness and an interest in seeing their texts in lively negotiation with their own and successive cultures.

Titles include:

Arthur F. Marotti (*editor*)
CATHOLICISM AND ANTI-CATHOLICISM IN EARLY MODERN
ENGLISH TEXTS

Mark Thornton Burnett
MASTERS AND SERVANTS IN ENGLISH RENAISSANCE DRAMA AND CULTURE
Authority and Obedience

The series Early Modern Literature in History is published in association with
the Renaissance Texts Research Centre at the University of Reading.

Early Modern Literature in History
Series Standing Order ISBN 978-0-333-71472-0
(*outside North America only*)

You can receive future titles in this series as they are published by placing a standing order.
Please contact your bookseller or, in case of difficulty, write to us at the address below with
your name and address, the title of the series and the ISBN quoted above.

Customer Services Department, Macmillan Distribution Ltd, Houndmills, Basingstoke,
Hampshire RG21 6XS, England

Betraying Our Selves

Forms of Self-Representation in Early Modern English Texts

Edited by

Henk Dragstra

Sheila Ottway

and

Helen Wilcox

First published in Great Britain 2000 by
MACMILLAN PRESS LTD
Houndmills, Basingstoke, Hampshire RG21 6XS and London
Companies and representatives throughout the world

A catalogue record for this book is available from the British Library.

ISBN 978-0-333-74029-3

First published in the United States of America 2000 by
ST. MARTIN'S PRESS, LLC,
Scholarly and Reference Division,
175 Fifth Avenue, New York, N.Y. 10010

ISBN 978-1-349-62849-0   ISBN 978-1-349-62847-6 (eBook)
DOI 10.1007/978-1-349-62847-6

Library of Congress Cataloging-in-Publication Data
Betraying our selves : forms of self-representation in early modern English texts /
edited by Henk Dragstra, Sheila Ottway, and Helen Wilcox.
p.   cm. — (Early modern literature in history)
Includes bibliographical references and index.

1. English literature — Early modern, 1500–1700 — History and criticism. 2. Self in
literature. 3. Autobiography. I. Dragstra, Henk, 1947– II. Ottway, Sheila, 1949–
III. Wilcox, Helen. IV. Series.

PR438.S45 B48   2000
820.9'003 — dc21

00–035262

Contents

Acknowledgements

The editors of this collection of essays gratefully acknowledge the help that they have received from a number of people and institutions.

For assistance with the conference at which a number of these essays were originally presented, 'All By Myself: The Representation of Individual Identity in Autobiographical Writings', held at the University of Groningen, The Netherlands, in November 1996, they would like to thank: the Royal Dutch Academy of Arts and Sciences, the Rudolf Agricola Institute for Research in the Humanities (Groningen), the Departments of English and Comparative Literature at the University of Groningen, and the following individuals: Carolyn Ayers, Menno de Leeuw, Gorus van Oordt, Duco van Oostrum and Marieke van Tol.

For assistance with the preparation of this volume, the editors wish to thank: Cedric Brown, Charmian Hearne, Julian Honer, Ruth Willats and Selma van der Ploeg, as well as, of course, the contributors themselves. They are also grateful to the Province of Groningen for permission to reproduce Herman Collenius's 'Allegorie op de Gerechtigheid' (1699) as jacket illustration.

For their support during both the conference and the editorial stages, the editors happily acknowledge the continuing tolerance and good humour of their colleagues, partners and families.

Notes on the Contributors

Cedric C. Brown, general editor of the series *Early Modern Literature in History*, is Professor of English and Head of Department at the University of Reading. He is author of two books on Milton, various essays on seventeenth-century poetry and entertainments, and has edited or co-edited various volumes including (with Arthur F. Marotti) *Texts and Cultural Change in Early Modern England* (1998). His current project, of which this material is a part, is a book about textual transmission and social functions of poetry texts.

Henk Dragstra teaches English literature at the University of Groningen, The Netherlands. His research interests include the history of popular culture and in particular the early modern street ballad. He is the editor of a forthcoming essay collection, *Beggars Description: Confronting Destitution in English-language Literature*.

Elspeth Graham is principal lecturer in literature and cultural history at Liverpool John Moores University. She co-edited *Her Own Life: Autobiographical Writings by Seventeenth-Century Englishwomen* (1989). She has published articles on Bunyan and Milton, feminist literary theory and seventeenth-century autobiography and women's writing, including a chapter entitled 'Women's Writing and the Self', in *Women and Literature in Britain, 1500–1700* (1996) edited by Helen Wilcox.

Peter Happé is currently Visiting Fellow at Southampton University, and a Research Associate of the Open University. He has recently published *John Bale*, an edition of Jonson's *The Devil is an Ass*, and *English Drama before Shakespeare*. He is working on a book on the English cycle plays and their continental analogues, and editing plays by Ben Jonson.

Elizabeth Heale has most recently published a study of early Tudor court poetry, *Wyatt, Surrey and Early Tudor* (1998). She is working on a book on autobiographical poetry and the sense of the self in sixteenth-century poetry for the Macmillan 'Early Modern Literature in History' series, edited by Cedric Brown. A revised second edition of her *The Faerie Queene: A Reader's Guide* is forthcoming.

Harald Hendrix is associate professor of comparative literature at Utrecht University. He works on late sixteenth- and early seventeenth-century Italian literature, and is the author of *Traiano Boccalini fra erudizione e polemica: Studi sulla fortuna e bibliografia critica* (1995). He has edited *The Search for a New Alphabet: Literary Studies in a Changing World* (1996) and, for the series Utrecht Renaissance Studies, a number of volumes, including *De vrouw in de Renaissance* (1994) and most recently *De grenzen van het lichaam* (1999).

Allan Ingram is professor of English at the University of Northumbria at Newcastle. His books include monographs on James Boswell, Swift and Pope. Most recently he has published *The Madhouse of Language: Writing and Reading Madness in the Eighteenth Century* (1991), and he is the editor of *Voices of Madness: Four Pamphlets, 1683–1796* (1997) and *Patterns of Madness in the Eighteenth Century: A Reader* (1998). He is currently working on representations of insanity in the eighteenth century.

Simon Meecham-Jones is currently tutoring in Cambridge. He has researched extensively on medieval literature, and is preparing a book on Chaucer's presentation of the natural world, and the light this sheds on his *Poetics*, as well as a new edition of the fifteenth-century treatise on book illustration, *The Crafte of Limning*.

Sheila Ottway recently gained her PhD from the University of Groningen, with a thesis entitled *Desiring Disencumbrance: The Representation of the Self in Autobiographical Writings by Seventeenth-Century Englishwomen*. She is currently living in Oxford.

W.A. Sessions is Regents' Professor of English at Georgia State University. He is the author of many works on Francis Bacon, and a critical biography of Henry Howard, Earl of Surrey (1999).

Helen Wilcox is professor of English literature at the University of Groningen, with teaching and research interests in a wide range of early modern literary forms and issues. Publications relevant to this volume include *Her Own Life: Autobiographical Writings by Seventeenth-Century Englishwomen* (1989, co-edited with Elspeth Graham, Hilary Hinds and Elaine Hobby), and *Women and Literature in Britain, 1500–1700* (1996).

Ramona Wray is lecturer in English literature at The Queen's University of Belfast. She is the co-editor of *Shakespeare and Ireland: History, Politics,*

Culture (1997) and of *Shakespeare, Film, Fin de Siècle* (forthcoming), and the author of *Women Writers of the Seventeenth Century* (forthcoming). She is currently working on *Women, Writing, Revolution: An Anthology of Writing by Women during the English Civil War.*

Marion Wynne-Davies is senior lecturer in English Literature at the University of Dundee. She has published widely on women writers in the early modern period. Her books include *Renaissance Drama by Women: Texts and Contexts* and *Readings in Renaissance Women's Drama: Criticism, History and Performance, 1594–1998*, both with S.P. Cerasano; and *Women Poets of the Renaissance.*

Introduction

Henk Dragstra, Sheila Ottway and Helen Wilcox

> None can accuse us, none can us betray,
> Unless our selves, our own selves will bewray.
> (Mary Wroth, *Love's Victory* III.ii.25–6)

This collection of essays is an exploration of written forms of self-representation in the period before the advent of autobiography as a recognised genre,[1] concentrating in particular on English texts from the sixteenth and seventeenth centuries. It does not aim to be a comprehensive pre-history of autobiography, but addresses a number of crucial questions for the study of the early modern period, and for our increasing understanding of autobiographical writing. There are fundamental matters of information to be gathered: what sorts of people, for example, attempted to represent their own selves in writing in this period, and what literary forms were the most accessible or appropriate? There is a range of interpretative questions to be considered, including those of influence and motivation, and of whether the self-representation was conscious or incidental. Underlying all of these are the thorny problems implicit in the term 'self-representation': what exactly does literary representation entail, and with what confidence can we use the word 'self' with reference to early modern writers?[2]

The essays are arranged in the chronological order of their subjects, since the questions addressed need to be considered in the context of historical and literary developments during the period. The main body of essays is framed by a prologue and epilogue, the former looking back to the work of the medieval poet John Gower in order to ask when poetic self-consciousness began, and the latter looking forward to probe the parallels between early modern and twentieth-century representations

1

of the suffering self. The eleven essays between these two chronological extremes consider the autobiographical tendency as it took shape in English texts from the sixteenth and seventeenth centuries, at the hands of writers as well known as Francis Bacon and as little known (hitherto) as the Derbyshire yeoman poet, Leonard Wheatcroft. Men and women, lowly and aristocratic, devout and secular, sane and insane, all feature in the lively discussions which follow.

Forms of Self-betrayal

The early modern writers whose work is the subject of this book have one thing in common: they have all 'betrayed' themselves through their texts. Mary Wroth's lines, quoted above and echoed in our title, serve to highlight the complexity of self-representation in the early modern era. The autobiographical writings discussed in the following essays are just as often to be found disguised within other forms of writing as in an obviously self-promoting mode. In this sense, early modern autobiographical texts may be seen as 'betraying' the nature of their writers, often unintentionally giving away their presence – or perhaps their secrets. We may, of course, consider this effect of betrayal to be a positive one; these early autobiographical texts allow the writer's identity to be observed, and frequently, if unconsciously, declare a distinctive character. However, betrayal may also have an actively negative impact, as suggested in Wroth's lines where the word is linked with accusation by others; a betrayal can be a misrepresentation or endangering of the self.

Autobiographical writing undoubtedly renders the author vulnerable and can be as much a threat as a liberation. Mary Wroth's preference for the rhyming word 'bewray', when speaking of what the group of women in her play *Love's Victory* might do by their conversation about themselves, is very telling. 'Bewray' was a consciously archaic term by the seventeenth century – largely replaced by 'betray' – but seems to have been particularly associated with divulging or disclosing one's own deeper being: as the Elizabethan Archbishop Sandys wrote in 1589, 'A man's speech or gesture will bewray his inner thoughts'.[3] In this collection we are concerned, as Mary Wroth's characters were, with the tensions between the shifting meanings of 'betrayal' and 'bewrayal' – the almost inseparable fear and fascination of self-revelation – in a variety of early modern texts.

The literary forms enabling self-expression in the sixteenth and seventeenth centuries came from a surprising range of traditions and genres. There are modes of self-representation to be found in the formality of the poetic line, as the essays of Simon Meecham-Jones on Gower, and Elizabeth Heale on lyric self-dramatisation, demonstrate. The dominant rhetoric of patronage, Harald Hendrix argues, allowed Elizabethan authors to construct themselves between the lines of the Italian-style courtly letter. Writers of ecclesiastical history (John Bale, discussed by Peter Happé) and of philosophy (Francis Bacon, discussed by W.A. Sessions) found a place and a use for self-representation in their apparently objective texts. Marion Wynne-Davies goes so far as to suggest that there is an allegorical code of self-identification in the Sidneian pastorals of Lady Mary Wroth, here hinting at an identity which is social and familial rather than individual. It is not only the autobiographical instinct which can be hidden within these indirect modes, but the very idea of the self may find its expression within a doctrinal, philosophical or social group, rather than in the post-enlightenment isolated individuality that we may be accustomed to expect.

Self-betrayal is therefore to be discerned in many indirect and hidden modes in early modern writing – through incidental revelation, secreted in allegory, or as a means to a personal, intellectual or doctrinal end. It is also rarely to be found in a clear or finished form. Cedric Brown's essay introduces the idea of autobiography taking the shape of a personal miscellany of texts, as in the case of those written by the Derbyshire yeoman poet, Leonard Wheatcroft. Ramona Wray explores the diametrically opposed versions of the life of Mary Rich, both written by the woman herself but reflecting different pasts – and different selves – given form by the contrasting modes of diary and narrative, and the varying needs of editors and readers. The sense that there is no single, definitive self or life to be recorded is also prominent in Helen Wilcox's discussion of the self-representations of women caught between material and spiritual priorities, and in Sheila Ottway's comparison of two memoirs by different writers, Anne Halkett and Joseph Bampfield, whose lives were closely linked but whose reconstruction of events are intriguingly at odds with one another. A further challenge to the idea of a straightforwardly inscribed life is delivered by Henk Dragstra's investigation of autobiographical traces in the semi-oral tradition of the broadside ballad. Our notion of that which defines autobiographical writing may need to shift not only to include indirect modes of self-expression, but also to contain self-representations at the fringes of the conventionally literary, or literate.

Earlier readings of seventeenth-century autobiographical writings have tended to stress the confessional narrative, as exemplified by John

Bunyan's *Grace Abounding to the Chief of Sinners* (1666), as the most readily available form of self-expression in this period.[4] Although it was undoubtedly important, particularly in the context of radical religious self-expression, the genre needs to be reassessed in at least two ways. First, as Allan Ingram suggests in his consideration of Hannah Allen's *Satan's Methods and Malice Baffled* (1683), the genre of exemplary confession could hide a variety of personal narratives – of suffering, of sexual or social need, even of madness – within its formulaic and didactic outlines. Second, once we recognise that autobiographical expression is to be found in an enormous variety of other seventeenth-century modes, such as the commonplace book, the collection of private meditations, the published miscellany or the ballad, then the narrative autobiography becomes merely the tip of the iceberg.

Models of Self-representation

The essays in this volume bear witness not only to a diversity of autobiographical forms in the late medieval and early modern period but also to a diversity of narrative models upon which autobiographers could base their textualised self-portraits. In discussing possible reasons for the notable rise in concern with the self in European culture of early modern times, the cultural historian Peter Burke has emphasised the increasing availability of models of selfhood that were the legacy of both Christian and pagan antiquity. In refuting the Burckhardtian vision of a nebulous spirit of individualism awakening in Renaissance Italy, Burke states that 'the rise of the autobiographical habit was not an inexplicable change in "spirit" but a chain reaction, in which certain texts awoke or restructured perceptions of the self, while these perceptions in turn created a demand for texts of this kind.'[5] The texts discussed in this volume are to be seen as links in this chain reaction, inasmuch as the authors of these texts were all involved in a shaping of their individual selfhood partly on the basis of pre-existing models that were privileged by contemporary culture. Such models often derive from the world of classical antiquity or the Bible, or from a combination of both. The spread of literacy and print culture in the early modern period obviously facilitated the propagation of both sacred and profane texts that provided suitable models for emulation. Whereas modern autobiographies are often characterised by a preoccupation with the author's uniqueness as an individual, autobiographical writings of early modern times are more concerned with identity in the original sense of the word: that is to say, the degree of 'sameness' or similarity between the author and his or her chosen model. In practice, autobiographical texts produced in the early modern period are remarkably diverse and all unique in their own way, as the essays in this volume testify; in theory, however, the authors concerned attempted, by and large, to inscribe their selfhood in accordance with culturally sanctioned patterns of exemplariness.

The influence of the world of classical antiquity on late medieval and early modern autobiographers is evident in several essays in this volume. Thus Simon Meecham-Jones draws attention in his essay to the way in which John Gower identifies himself in his *Confessio Amantis* with classical authors like Virgil and Ovid; moreover, not only the poet but also the poem itself follows classical model, its Middle English text being accompanied by a Latin apparatus of scholarly notes, provided by the author. According to Meecham-Jones, Gower's textual self as portrayed in the *Confessio Amantis* embodies the values not only of classical antiquity but also those of Christianity: Gower defines his poetic practice in accordance with Christian notions of humility, and presents his poem as an apologia for having dedicated himself to a life of study. Such a blend of classical and Christian traditions is also evident in the autobiographical passages in the writings of Francis Bacon, as discussed in the essay in this volume by W.A. Sessions. Thus it is shown how in such works as the *Instauratio Magna* Bacon assumes the role of the classical orator of antiquity, though with appropriate Christian humility; moreover, according to Sessions Bacon portrays himself in this work as 'a saint-like figure giving birth through his suffering to the modern world' (p. 95). If such suffering is intellectual, rather than physical, being concerned with a radically new way of thinking, the association with the Christian idea of suffering as an ennobling experience is clearly apparent. Sessions argues that the new kind of subjectivity that is evident in Bacon's writings originates in the writings of Erasmus, notably in the discursive mode of his *Enchyridion militis Christiani* (Handbook of a Christian Soldier) of 1501; as Sessions points out, what is striking about this new subjectivity is that the relationship of self and Other is of fundamental importance.

The use of biblical models is also conspicuous in the autobiographical writings discussed in this volume of essays. The medieval ideal of the imitation of Christ finds resonance in the captivity narrative of the seventeenth-century Quakers Katherine Evans and Sarah Cheevers, as explained in Elspeth Graham's contribution to this volume. Imprisoned on Malta by the Italian Inquisition, Evans and Cheevers found succour in their ability to identify with the suffering Christ. Graham's analysis of Evans' and Cheevers' narrative is used to demonstrate her contention that it is through suffering that selfhood comes into being. Apart from the figure of Christ, the apostle Paul often functions as a model for early modern autobiographical writings. Peter Happé's essay in this volume shows how John Bale identifies closely with the figure of St Paul in his autobiographical *Vocacyon*, not only by recognising parallels between his own life and that of the apostle, but also in imitating the rhetoric of the New Testament Epistles written by St Paul to the early Christians. Happé points out that in another work by Bale, his *Anglorum Heliades*, the author gives a brief account of his own life in which he sees his conversion to Protestantism as a key event in his life. It is noteworthy that both the biblical account of the conversion of St Paul, on the road to Damascus, and the autobiographical account of the conversion of St Augustine of Hippo, as described in his *Confessions*, constituted narrative models that were of fundamental importance to early modern Christian autobiographers, especially Protestants. For such autobiographers, the conversion narrative provided a means of structuring one's life story in such a way that one could find meaning in the events of one's past. The chronological pattern of a period of religious doubt and dejection, followed by a conversion experience bringing spiritual regeneration, is evident in many autobiographical writings of early modern times, for example the personal narrative of Hannah Allen, discussed in this volume by Allan Ingram. As Ingram explains, the pattern of religious interpretation that Allen succeeded in imposing upon her past life enabled her to distance herself definitively from her former state of severe mental instability: in writing her own conversion narrative she was able to create a textual self that affirms her recovery from insanity.

While classical and biblical models provided early modern autobiographers with a template for the shaping of their individual identity, the way in which these authors inscribed their individuality in their self-writings often bears witness to a strong sense of belonging to a community or being devoted to a common cause. The kind of community concerned could be sectarian, as in the case of the Quaker missionaries Evans and Cheevers, or scholarly, as in the case of Francis Bacon. The autobiographies of Anne Halkett and Joseph Bampfield, compared and contrasted in the essay in this volume by Sheila Ottway, each testify to the author's sense of commitment to a political cause, namely that of the Royalists during the English Civil War. While Halkett remained pro-Royalist throughout her life, Bampfield's autobiographical *Apology* was written partly to justify the shift in his political allegiance which he was forced to make after being shamefully mistreated by his royal master,

Charles Stuart. Towards the other end of the social scale, the yeoman poet Leonard Wheatcroft, whose autobiography and courtship narrative are discussed in the essay in this volume by Cedric Brown, presents himself in his self-writings as a local laureate and newsmonger who served as a figure of authority in the rural community to which he belonged. Wheatcroft's somewhat comical self-celebration as rustic bard and as instructor-wit to the youth of Derbyshire is indicative of his strong sense of rootedness in the local community; as Brown points out, it is precisely towards his fellow-countrymen that Wheatcroft directs his self-celebration in his semi-fictional autobiographical texts.

If the definition of early modern individual identity is often strongly influenced by political or social affiliation, it is noteworthy that family ties and gender are also important elements involved in the shaping of the autobiographical self. Marion Wynne-Davies' essay in this volume on Mary Wroth demonstrates how Wroth took part in the highly sophisticated textual game-playing that went on in the Sidney/Herbert family circle to which she belonged, by recreating herself as a fictional character in her own writings. Wynne-Davies argues that in her play *Love's Victory* Wroth uses the autobiographical mode in portraying not only herself as a fictional character but also several noblewomen of her acquaintance; in this way Wroth celebrates the close ties of female friendship that existed in real life within a coterie of noblewomen, to which she herself belonged. Many of the women referred to in this volume found themselves defined against contemporary expectations of the female role in marriage, as defined in romance and conduct literature as well as the Bible, but forged their own identity both with and counter to such models. As the lives of Mary Rich (discussed by Ramona Wray), Martha Moulsworth and Mary Carleton (among those considered by Helen Wilcox) all in their varied ways make fascinatingly clear, gendered identity was a fluid process, always forming itself in negotiation with existing models and norms.

Motivations for Self-writing

When we turn to consider *why* the subjects of the essays in this volume wrote in ways that 'betrayed' (or 'bewrayed') their own selves, it immediately becomes clear that there was little or no straightforward self-portrayal motivated by a new sense of personal worth, let alone any triumphant construction of the self as 'universal man', such as that of Leon Battista Alberti. Burckhardt's claims may well have applied in Renaissance Italy, in courtly surroundings, among aristocrats, of the male sex; but our material certainly does not suggest it did in Britain. Only Leonard Wheatcroft, as late as the second half of the seventeenth century, presents himself uncomplicatedly as God's gift to the reader. He had a capacity, says Cedric Brown, 'for laureate self-display and the grand gesture' (p. 122). Without speculating through what channels this upstart might have assimilated such secondhand courtly views of the self, we cannot fail to see in Wheatcroft's texts the elements of *persona* and of *manera*, in short, of self-promotion. There is not much divulging of inwardness in his writing: he only shows off what he is proud of, including the application of accomplishments like the 'gentlemanly' wooing of his wife; and his 'lyrical' accoutrements are sometimes borrowed plumes. But his work is exceptional, both in its self-congratulatory attitude and in the openness with which it is exhibited in texts.

In several cases considered in this volume, self-writing is not presented as an aim in itself: autobiographical elements are billeted upon texts that profess to deal with subjects of public interest. Francis Bacon offers an interesting paradox in this respect: as W.A. Sessions puts it, he uses 'self-aggrandisement to demonstrate himself . . . as a model of humility' (p. 95). Bacon found a justification for self-exhibition in the discursive purpose of his text, which demanded this approach – or so he claimed. Simon Meecham-Jones finds John Gower making a 'playful show of assumed humility' in his *Confessio Amantis* in order to exhibit 'the accumulation of a lifetime's wisdom' and 'to stake his claim to be classed with the immortals' (pp. 23, 28). In Johan Bale's *Vocacyon*, an autobiographical account is sandwiched between solid slabs of religious rhetoric, which provide the excuse as well as the framework for his personal narrative.

But even where texts reviewed in our collection are wholly autobiographical, most authors by no means present a grand picture of themselves; and in most cases their status in life would have made such ambition impossible, or ludicrous. Nor is humility in such texts necessarily a 'playful show'. Several of the writers in question were women, and as Helen Wilcox points out, their dependent social and economic position was etched deeply into their sense of self. Most of the male authors are of low or indifferent birth; indeed, the material shows a remarkable emphasis on ordeals caused by disfavour, deprivation, low life, imprisonment, exile, madness and public execution (for a contemporary catalogue of 'parells', see Bale's account, pp. 49–50). If these texts represent in any way the beginnings of English autobiography, we may justly call its origins humble.

So the 'Italian' approach to autobiography will only take us part of the way into our material. Motivations for writing the self, or for having one's self written, may be entirely negative, as Henk Dragstra shows in his analysis of 'Hanging Ballads', which had the overt purpose of humiliating, indeed of annihilating, the subject's individuality. As several essays attest, early autobiographies are more often the product of suffering and humiliation than of pride and success. But this 'English' approach to the question of self-representation has its own dangers. It is tempting to present the multifarious texts that contributors to the present collection have been able to glean from a hitherto seemingly barren field as 'proto-autobiography'. And since the period we are concerned with is now commonly known as 'early modern' rather than 'Renaissance', we naturally tend to see the works and authors discussed from a present-day retrospective angle. Put these two factors together, and the material presented appears as budding or perhaps embryonic autobiography, reflecting a nascent, incipient, or even ur-modernity of spirit.

As long as we are thinking in terms of genre, this view is defensible. People have lived their lives from time immemorial without writing about them; evidently the urge to render one's life as a written or printed text is not of all times (though the tendency to see one's life as a story, myth or dramatic role probably is). If the history of autobiography is the history of that urge to self-inscription, then the sixteenth and seventeenth centuries in Western Europe can justifiably be described, in a biographical metaphor, as the infancy of autobiography. This is an exciting vision because it suggests a link of consciousness between *us* and *them*, promising an easy accessibility to early modern identities, to which autobiographical texts provide the key. Without qualifications and reservations, however, this approach can put us completely out of touch with the texts as they were written and read in their own day.

When we examine the 'beginnings of autobiography', there is no denying that each individual text is, besides a postmortem on a portion of its author's life, the culmination of a complex of motives rooted in the contemporary situation. Peter Happé finds in Johan Bale's *Vocacyon*, for example, an 'underlying ... polemical purpose' (p. 45), and Harald Hendrix perceives early modern self-presentation as a product of rhetorical negotiation. Elizabeth Heale explores the influence of the market for print culture on mid-sixteenth-century lyric self-expression, while Elspeth Graham demonstrates how the self-definition of Quakers tended to have a consciously 'oppositional element' (p. 203). Even an innovative and forward-looking genius like Francis Bacon did not generate his scientific achievements by groping blindly into an unseen future. Like

Columbus, he was following a vision, not for its own sake but out of dissatisfaction with the present, which was too much encumbered with the past; and the identity that he reveals in the process of writing is, as W.A. Sessions puts it, 'a self in the midst of its own terrifying history' (p. 94).

If a comparison of early forms of autobiography to buds or to infants is justified, it is not by their tenderness, but by their stubborn insistence on coming into being. The self at the centre of these autobiographical texts is a self in the midst of pressure, a self from which utterances are forcefully squeezed; in many cases, this is unfortunately not just a metaphor. Examples of subjects under physical torture have already been mentioned; but mental torture is just as important a factor. Allan Ingram presents prime examples in Hannah Allen and Samuel Bruckshaw, both beset by bouts of insanity and by the social ostracism that these inflicted upon them; Ramona Wray shows how Mary Rich was torn between two completely different emotions towards her husband. Joseph Bampfield experienced his state of disfavour and exile as a tomb, and his writing is at times 'a veritable cry from the heart' (Ottway, p. 143).

But in spite of the mental anguish, these latter writings are not mere cries of pain; their emergence into writing as *voces clamantium* was caused by social as well as private purposes. The texts written by Bampfield, Allen and Bruckshaw have all come to us in the form of pamphlets in which they seek vindication or rehabilitation; in this sense, they are forms of self-promotion as much as Wheatcroft's writings are. Indeed the need or wish to project a positive self can be said to form the wellspring of any real autobiography; the exception of 'Hanging Ballads' only helps to highlight this fact. As the products of the printing press gained more and more foothold and familiarity in people's daily lives, it became correspondingly more natural for oppressed individuals to avail themselves of that medium. The autobiographical pamphlets mentioned might not have been written at all if the writers' need to improve their damaged public image had not provided a strong motive for putting them into circulation.

Anne Halkett's memoirs were never intended for print; yet, or perhaps because of this, they are very attractively written, conveying a sense of intimacy, subjectivity and domesticity. This is a kind of autobiography that modern readers find easy to appreciate; all the more reason to be on our guard. As Sheila Ottway shows, Halkett was no different from her former lover Bampfield in her need for self-vindication; indeed there is an element of self-celebration in her memoirs as well. The reason why her text was not published until many years after her death was not, perhaps, that the author was too modest, but that publication would have been too risky. It was wiser to let the memoirs circulate in a private circle only.

The fact that some texts were emphatically private does not mean that they constitute pure self-expression or an uncomplicated narration of events. Indeed, veiled autobiography can become an art form in its own right, as the literary games played at Penshurst demonstrate. Here, as Marion Wynne-Davis demonstrates, the privacy of the 'autobiography' is that of a group, not an individual: a small circle where 'None can accuse us, none can us betray' (p. 81). Nevertheless, this partial secrecy is no guarantee of artless frankness; self-betrayal is confidently encoded in the pastoral idiom with a conscious play on the overlapping patterns of fact and fiction. Even privacy can have a 'show' element. And here 'Italian' aspects of British autobiography reassert themselves. When we watch Leonard Wheatcroft's public display of accomplishment and experience, the ostentatiousness is so obvious as to be almost comical. We cannot resist the impulse to guess at the private self hidden behind his rustic mannerism; indeed, we cannot help seeing the amateurishness with which he performs his rituals. In Mary Wroth's works, the need for privacy, particularly for the female self, first springs into view; but, taking the form of game-playing and play-acting, her text also involved display. The rhetorics of the two individuals are undeniably different in tone; but in purpose, they are not unalike.

In the writing of early modern autobiography, the apparently contradictory modes of keeping silent and speaking out, humility and ostentation, are equally valid ways to project an appropriate image or establish a favourable reputation. This fact is easy to overlook if we take for granted autobiography as an autonomous genre and self-expression as an activity for its own sake. Early modern writings of the self, various and seemingly haphazard as they are, confront us excitingly with the centrality of motives that we would perhaps rather see as extraneous.

Images of Self-representation

The essays in this volume are rich in images used by early modern writers, whether deliberately or indirectly, to depict the particularity or exemplary quality of an individual life. One of the pleasures of reading this volume, we trust, will be the encounter with these individuals through the medium of language. All the authors, however, were engaged, as we have already suggested in this introduction, in complex processes of self-representation; their self-writing was mediated by literary forms and available models of selfhood, and driven by motives embedded in circumstance as well as choice.

The cluster of possibilities implicit in the idea of autobiographical writing in the early modern period may be perceived in the images in the illustration on the cover of this book. Painted by Herman Collenius in 1699 and now hanging in the Provinciehuis of Groningen in the Netherlands, this 'Allegorie op de Gerechtigheid' (Allegory of Justice) implies a great deal about the mixture of self-justification, self-betrayal, judgment and truth which informs early modern autobiographical writing. Seated on a throne at the centre of the painting is Justice, surrounded by, on the left, naked Truth and willing Innocence, and, on the right, careful Prudence and active Wisdom unmasking Deceit. The emblems used by Collenius in his allegorical painting are vividly suggestive of the elements combined in autobiographical writing. The mirror, held up by Prudence, is a classic image of self-knowledge as well as self-recognition. Innocence offers up a heart, icon of sincerity and the seat of individual life and affection, but stands with a lamb, reminding us of the association of both hearts and lambs with the sacrificial victim. Truth, meanwhile, stands unashamedly naked, hiding nothing of her self, whereas the fraudulent old woman is shown having her true face revealed from beneath a mask. The practice of autobiographical writing is never so clear-cut, of course; masks are not always deceits, and a rhetoric of sincerity is a sure mark of neither innocence nor honesty. The sword held by Justice, on the other hand, reminds us of the need for self-defence which often drives autobiographical writing, and her scales recall the attempt towards – if not always the achievement of – a balance between 'betrayal' and 'bewrayal', revelation and confession, in self-representation. Above all, Collenius's painting also contains a book – held by Truth – as witness to the events of a life.

The early modern texts discussed in our volume all participated in the commitment of lives to the witness of the written or printed page, and function at the meeting point of word and deed, life written and life lived. They await your analysis and – to continue the allegory – your judgment. Where these texts lie in the spectrum of possibilities between mirrors and masks, swords and sacrifices, we leave for you to decide.

Notes

1 The term 'autobiography' was first used in English in 1807 by Robert Southey, and the widespread recognition of autobiography as a literary genre

worthy of inclusion in, for example, university curricula is a late twentieth-century phenomenon. See James Goodwin, *Autobiography: The Self Made Text* (New York: Twayne, 1993), and James Olney, *Studies in Autobiography* (New York: Oxford University Press, 1988). However, as Sheila Ottway has demonstrated in *Desiring Disencumbrance: The Representation of the Self in Autobiographical Writings by Seventeenth-Century Englishwomen* (University of Groningen PhD thesis, 1998) pp. 8–10, the practice of autobiography was certainly described in the seventeenth century, by Margaret Cavendish in 1664 and Roger North in the 1690s.

2 Among a number of recent discussions of these points, see *The Making of Sixteenth Century Identity*, ed. Amanda Piesse (Manchester: Manchester University Press, 1999); *Rewriting the Self: Histories from the Renaissance to the Present*, ed. Roy Porter (London: Routledge, 1997); Michael Mascuch, *Origins of the Individualist Self: Autobiography and Self-Identity in England, 1591–1791* (Cambridge: Polity Press, 1997); and John Martin, 'Inventing Sincerity, Refashioning Prudence: The Discovery of the Individual in Renaissance Europe', *American Historical Review* 102.5 (1997) 1309–42, an essay which deals, among other topics, with the influential but disputed work by Stephen Greenblatt, *Renaissance Self-Fashioning: From More to Shakespeare* (Chicago: Chicago University Press, 1980).

3 Archbishop Sandys, *Sermons* (London, 1841) p. 395.

4 See, for example, Paul Delaney, *British Autobiography in the Seventeenth Century* (London: Routledge and Kegan Paul, 1969), and Margaret Bottrall, *Every Man a Phoenix: Studies in Seventeenth-Century Autobiography* (London: John Murray, 1958).

5 Peter Burke, 'Representations of the Self from Petrarch to Descartes', in *Rewriting the Self*, op. cit., pp. 17–28; the quotation is from p. 27. The idea of Renaissance individualism originated with the publication, in 1860, of Jacob Burckhardt's *Die Kultur der Renaissance in Italien* [*The Civilisation of the Renaissance in Italy*, translated by S.C.G. Middlemore (Oxford, 1945)].

1
Prologue: the Poet as Subject: Literary Self-consciousness in Gower's *Confessio Amantis*

Simon Meecham-Jones

Just as the title 'the Middle Ages' was invented by later generations to mark out a perceived distance from past consciousness, so critics have tended to accept that certain modes of expression and certain modes of thought could not have been present in literature before the development of 'modern' conceptions of the unique value of human personality. Yet, even if it must be accepted that (surviving) medieval conceptions of authorship provide little or no encouragement for writers to present accounts of their own experience as being of intrinsic value, it would be surprising if a mode so central to oral literature had ever disappeared, or been banished, from the stylistic models of written culture. Nor is it possible to reconcile Foucault's confident assertion that

> The idea that from one's own life one can make a work of art is an idea which was undoubtedly foreign to the Middle Ages and which reappears at the moment of the Renaissance[1]

with the creation of texts as intent on the exemplary value of lived experience as Abelard's *Historia Calamitatum*, *The Boke of Margery Kempe* or Mandeville's *Travels*.

The reticence of medieval authors in making use of the autobiographical style is better understood as expressing an anxiety at appearing to set their works in competition with the 'authoritative' texts of the revered literary past. From both biblical and classical practice, medieval authors were inspired by, and burdened with, the example of texts purporting to be autobiographical in style. Throughout the Middle Ages a series of key texts reappear as influences – sometimes directly referred to, sometimes recalled in the structure or diction of medieval texts. From the biblical tradition, medieval writers struggled to achieve the assimilation of models from The Book of Job, The Book of Jonah, the Prophetic Books (and in particular Isaiah), The Revelation of Saint John and the lyric expression of the Psalms within a personal voice at least nominally compatible with Christian notions of humility. The paramount model for Christian autobiography, though, was to be found in the intensely personal expression of St Paul, and medieval writers were forced to reckon with the rebuke to literary ambition his writings posed. In the neo-Latin lyric poetry of the twelfth century, the influence of St Paul was set against reminiscences of an alternative style – from the writings of Juvenal, Virgil (whose *Eclogues* and *Georgics* were read as containing autobiographical material) and pre-eminently Horace, a writer whose *Urbanitas* seemed to epitomise a psychic security unachievable by Christian authors disturbed both by the condition of their immortal soul and by the relation of their work to the authoritative texts of the past. The suitability of the autobiographical mode to provide a matrix within which 'authoritative' models could be appropriated is demonstrated clearly in the emergence of a self-consciously 'literary' tradition in the English language in the second half of the fourteenth century. Poems such as *Piers Plowman*, the alliterative *Pearl*, Chaucer's *House of Fame*, Thomas Usk's *Testament of Love* and John Gower's *Confessio Amantis* reveal contrasted and conceptually intricate stratagems through which the conventions of autobiographical writing could be employed to negotiate the difficulties of writing in a language not previously recognised as a suitable medium for literary endeavour, while showing proper respect for the practice of past 'auctors'. In each poem, it is the poet's recognition of the impossibility of language accurately representing 'real life' which permits the exploitation of autobiographical styles. As in so much medieval writing, it is the perceived artificiality of the autobiographical mode, and the author and audience's shared awareness of the literary nature of the convention, which guarantees its value to authors.

Where Chaucer's evocation of the autobiographical mode is generally occluded and consciously misleading, in the closing tableau of Gower's *Confessio Amantis* the reader is apparently allowed to witness a scene of autobiographical self-discovery, which proves to be an element in the poet's attempt to construct a validation of his own work. Every writer risks being judged within the expectations derived from one memorable component of their work, and the vituperative energy with which Gower castigated the vices of his society has obscured the maturing of his aesthetic evidenced in his exploration of a more self-questioning critical intelligence in the conception and execution of his most substantial work in English, the *Confessio Amantis*. Whereas Gower's major works in Latin and French had exploited the narrative potentialities of the poet figured as an historical witness (however partial) passing judgment on the events of his time, in the *Confessio Amantis* the status of the poet is no longer unproblematic, but must be accounted for and justified. The presumed consonance of the poetic voice *within* the text and its creator is dissolved by the artful manipulation of the autobiographical mode which paradoxically draws the reader's attention to the inevitable fictionality of all literature. The *Confessio Amantis* is most memorable not for the eloquence of Gower's style, or the vividness of its imagery, but for the subtlety with which Gower constructs a literary myth which enables him to reflect on his art and question its value.

Gower's perceived need at this stage of his career to attempt an examination of, and monument to, his literary aspirations is evidenced in the extended scope of the *Confessio* – by any standards an ambitious structure, enriched by the supplementary apparatus of an exhaustive sequence of Latin notes, generally attributed to Gower's hand, and which might have been considered more suitable for an edition of a classical work. In the attachment of a Latin apparatus to the *Confessio* Gower manages to assert simultaneously a primary distance between the representation of the figure of the 'poet' within the narrative and the historical poet and, more ambitiously, the literary dignity of the work on which he has engaged.

The first of these purposes is effected early in the course of the work, by means of a Latin headnote:

> Hic quasi in persona aliorum, alligat, fingens se auctor esse Amantem, varias huius libri distinccionibus per singula scribere proponit.

> [From here on the author, feigning himself to be a lover, as if in the person of those whom Love constrains, intends to write about their various passions one by one in the various sections of the book.][2]

But the hint is unnecessary, particularly since Gower is to achieve through the figure of 'Amans' a balance of sympathy and disengagement more complicated and subtle than that predicted in the headnote, with its emphasis on the pretence necessary for the poet to experience the emotions of love – 'fingens se auctor esse Amantem'.

None the less, the distinction highlighted in the headnote describes a conscious taking of control over the material of the work by a poetic figure not constrained by the force of love, and therefore able to achieve a disinterested analysis of the subject. In this way, Gower strives to establish the poet in a position simultaneously *within* and *outside* the texture of his poem. It is a sophisticated manoeuvre, through which the poet is able to license and restrict his ability, and thereby that of the reader, to identify with the predicaments of his characters. The structure of the work, with its integral but accompanying commentary and its concluding pantomime of *apparent* self-revelation, becomes implicated in the interpretation of individual episodes of the work, in a manner which casts doubt on judgments of Gower as an unreflective practitioner of his craft.[3]

The controlling scheme of the Seven Deadly Sins is presented as a classification of the particular nature of the distinctive sins of love, but whereas Gower's use of this traditional pattern in his earlier works had achieved a didactic clarity in the use of traditional *exempla*, it proves less certain when refracted within the overlapping levels of authorial presence enacted in the structure of the *Confessio*. Gower proves to have created a scheme that cannot avoid being ambiguous, involving as it does an attempted fusion of terms from two potentially antagonistic traditions of moral analysis.

An extreme example of the difficulties inherent in the scheme of the Seven Deadly Sins of Lovers is revealed, for example, in Gower's perhaps unexpected decision to adapt material from Ovid's *Heroides* in his retelling of the story of Canace and Machaire. The moral centre of this account of fraternal incest is surprisingly opaque, particularly since Gower is at some pains to show how the lovers' sin is caused by the external force of the Will of Love:

> Whan thei were in a prive place,
> Cupide bad hem ferst to kesse'
> And after sche which is Maistresse
> In kinde and techeth every lif
> Withoute laws positif,
> Of which sche takth nomaner charge,
> Bot kepth hire lawes al at large,
> Nature, tok hem into lore
> And tawht hem so, that overmore
> Sche hath hem in such wise daunted,
> That thei were, as who seith, enchaunted.
> (III 168–78)[4]

But while the lovers are represented as the victims of a force over which they have no control, it is their 'virtuous' father who is stirred to anger, the sin exemplified by this tale:

> He wolde noght his herte change
> To be benigne and favorable
> To love, bot unmerciable
> Between the wawe of wod and wroth.
>
> (III 214–17)

Although the lovers' conduct had inspired this exemplified sin, Gower appears to introduce an element of qualification into his account of their transgression, questioning the lovers' *intention* to sin, in a manner consonant with the challenging ethical analysis of earlier moral philosophers such as Abelard:

> Sicut autem non idem est velle quod voluntatem implere, ita non idem est peccare quod peccatum perficere. Illud quippe de consensu animi quo peccamus, hoc de effectu operationis est accipiendum, cum videlicet illud in quo prius consensimus opere implemus.[5]

> [Just as, indeed, to will and to fulfil the will are not the same, so to sin and to perform the sin are not the same. We should understand the former to relate to the consent of the mind by which we sin, the latter to the performance of the action when we fulfil in a deed what we have previously consented to.]

At the same time he condemns, through the narrative voice of the Confessor, the immoderate response of the father:

> Bot for al that he was to wyte,
> Thurgh his sodein Malencolie
> To do so gret a felonie.
>
> (III 333–6)

In choosing a pre-Christian tale, Gower can colour the text with the possibility that a Christian father, mindful of the need to forgive others their trespasses, might have been less 'unmerciable' in condemning the transgressions of his children, without undermining the prohibition on incest, which can be presumed to be so generally recognised as not to require restatement. The curious mixture of sympathy and judgment

Gower evokes in his handling of the difficult material of the story of Canace aptly characterises the necessarily hybrid note of a work which is constantly seeking to reconcile the opposing claims of two incompatible systems of values. It should not be thought, though, that this incompatibility reveals a failure of synthesising imagination on Gower's part. Rather, the very rigidity of the moral framework which acts as a skeleton for the poem's structure allows Gower to aim at an ambitious, if not unfettered, universality. It might have surprised Chaucer's Man of Law above all things to find the chronicler of such 'abhominations'[6] characterised as 'almost alone . . . in being a poet perfectly wellbred'[7] but he would surely have recognised Gower's claim to be considered an exhaustive chronicler of the consequences of love. The rigour incorporated through the schema of the Seven Deadly Sins is of crucial importance in allowing this examination, since it enacts in the poem's structure a constant reminder of the tenets of Christian teaching, while at the same time guaranteeing the moral orthodoxy of the poem. It is only within such an inescapably devout Christian framework that Gower can achieve the literary freedom to present potentially serious moral problems with a degree of flippancy or disengagement.

But the meaning of the 'ingeniously won moral freedom'[8] that the poem's structure allows Gower to achieve lies less in the understanding he is able to grant his sinners than in the sheer variety of situations his writing encompasses. Rather than claiming an especial value for the acuity, or divine inspiration, of his insights, Gower seeks to prove the value of his poem through the inclusiveness of its scope. It would be to misinterpret Gower to imagine that where there is understanding of human failings in his poetry, there is no condemnation, but the poem's narrative structure diffracts judgment in unexpected and personal directions. The poet's willingness to expand the permitted subject area of his verse into the territory of the pre-Christian masters, and the recognition of the different forces to which lovers are subject, adds substance to the poet's ability to encompass every element of this defining aspect of human experience within his poem, as if Gower aspires to be recognised as an encyclopaedist of Love. The bulk of the poem is, therefore, to be regarded as a guarantee of its quality, in so far as it witnesses Gower's assiduous garnering of material from the sources of inherited wisdom, as described in the poem's opening lines:

> Of hem that writen ous tofore
> The bokes duelle, and we therfore
> Ben tawht of that was write tho:
> (Prol. 1–3)

Gower's adventurous decision to engage with the potentially refractory subject of love is perhaps best understood as a homage to the texts of classical 'auctors', in whose poetry the consideration of love had been granted such especial prominence, and whose reputation and performance had helped to define Gower's conception of the nature of poetry. In a mischievous parody of the predisposition of audiences to read lyric poetry as presenting an accurate record of 'real' events, both Virgil and Ovid are depicted in their roles as lover, at the close of the poem:

> And ek Virgile of aqueintance
> I sih, wher he the Maiden preide,
> ...
> So dede Ovide the Poete.
>> (VIII 2714–15, 19)

and Gower points out the resemblance between the figures of these ancient poets and the poetic persona of Gower encountered in the *Confessio*:

> I thoghte thanne how love is swete,
> Which hath so wise men reclamed,
> And was miself the lasse aschamed,
> Or forto lese or forto winne
> In the meschief that I was inne:
>> (VIII 2720–4)

This identification of Gower's poetic self-image within the heritage of the classical tradition reflects a significant shift of emphasis in his career. In both his previous major works, the French *Mirour De L'Omme* and the Latin *Vox Clamantis*, Gower emphasises the role of the poet as a critic of human vices, in the figure of a prophet crying in the wilderness against the evils of the day.[9] In the *Confessio* Gower signals his reinvention of the poetic voice – no longer claiming the authoritative but exposed position of a conduit of divine Truth like Isaiah or Ezekiel. In a playful show of assumed humility, Gower opens Book I by marking out his distance, both from the prophetic mode and from Dantean pronouncements of literary ambition:

> I may noght strecche up to the hevene
> Min hand, ne setten al in evene
> This world, which evere is in balance:

> It stant noght in my sufficance
> So grete thinges to compasse,
>
> (I 1–5)

Instead, the poet turns his attention to a subject which presumably does lie within his own perception of his poetic competence:

> Bot I mot lete it overpasse
> And treten upon othre thinges.
> Forthi the Stile of my writinges
> Fro this day forth I thenke change
> And speke of thing is noght so strange,
> Which every kinde hath upon honde,
> And wherupon the world mot stonde,
> And hath don sithen it began,
> And schal whil ther is any man;
> And that is love, of which I mene
> To trete, . . .
>
> (I 6–16)

Though the nature of love is unruly and unpredictable,

> In which ther can noman him reule,
> For loves lawe is out of reule,
>
> (I 17–18)

it is only through the mediation of literature that its distinctive quality can be grasped – reinforcing Gower's conviction of the inextricable bond that links love and literature, which can be seen to provide the rationale for the *Confessio*'s structure.

The comprehensiveness of Gower's efforts to anatomise the diversity of the experience of love is of central importance to any consideration of the issue of poetic self-consciousness in the work, a theme inevitably linked to the scale, scope and inclusiveness of Gower's poem. The element of self-reflection in the poem is expressed through the complex relationship between the sections of the work and the relationship of the poet to the narrative voice of the poetic persona within the poem. Although the stories are presented within the fabric of the poem as the exempla narrated by Genius, the reader never forgets that they are ultimately the work of Gower himself, the sum of his observation and reading on the subject of love. Yet the meaning of the poem is lost if the reader splits the *Confessio* into a collection of mythological stories with optional trimmings, as C.S. Lewis seems to do:

> A consideration of the *Confessio Amantis* falls naturally into three divisions – the tales . . . the didactic passages and the love allegory in which all the rest are set.[10]

In fact, there can be no division, for the encounter of the poet's persona with Venus initiates the sequence of events which enables Gower to reflect on his performance as poet. The key is implicit in the title of the work, *The Lover's Confession*, for at an obvious level, the poet as lover has nothing to confess but the failure of his love. The appearance of Venus enables Gower to explain that seeming failure, and to do so, like Henryson in the *Testament of Cresseid*,[11] through the evocative image of a mirror:

> And forthi withal sche tok me tho
> A wonder Mirour forto holde,
> In which sche bad me to beholde
> And taken hiede of that I syne;
> (VIII 2820–3)

The comparison with Henryson is illuminating, for whereas in the *Testament*, a fictional character is brought to self-knowledge through the imagery of a mirror, in Gower's vision, the poet is forced to confront himself and render account for how he has spent his maturity. In Gower's work, the image is the more powerful for the recollection it brings of the organising imagery of his previous poem, the *Mirour de L'Omme*. The poet who previously sought to raise a mirror to all mankind is now forced to see himself without pretence, and in this moment of reckoning, Gower creates a visual image which serves to focus his summing up of his own career.

Through the fixed image of the mirror's revealing reflection, Gower is able to introduce into his narrative the theme of the passing of Time, which underlies the account of world history in the Prologue, but which the accumulation of tales has tended to obscure. But in the course of the poem, Gower's awareness of the passing of Time has become both more 'personal' and more potentially tragic. Like Rochester's inconstant lover he has been made aware of the brevity of the moment allowed to love:

> Then talk not of inconstancy,
> False hearts, and broken vows;

> If I, by miracle, can be
> This lifelong minute true to thee,
> 'Tis all that Heaven allows.[12]

For Gower also the moment has passed, but not through inconstancy. Instead, Venus' decision to call the poet by his name reminds us of the consequences of his work as poet. In the course of the poem, the poet has aged; after more than 30,000 lines, the reader has aged also. But it is the very minuteness of Gower's consideration of Love, and his dedication to the cumulation of his scholarship, which has left the poet an observer now unable to enjoy the fruits of love. It is as if Gower has taken from Chaucer the figure of a writer inexperienced in the practice of love[13] and extended it to demonstrate the inescapable impossibility of combining the active world of experience and the contemplative life of the author. He, the poet fated to be passive, has acquired the understanding for lack of which the figures in his exempla are destroyed. The poem is in that sense a confession, an apologia for the poet having devoted his life to study, when his earlier poems have shown a greater interest in, and commitment to, the active world of politics and social change.

It is at this point that Gower's distinctive conception of the nature of authorship at this stage of his career is revealed – not as the achievement of fine phrases or inspired imagery but in the accumulation of a lifetime's wisdom. For this reason it is imperative that the (apparently autobiographical) image of the poet appears within the text, since in this conception of poetry it is not possible for the author to achieve a separation from the valuation of his text.

It is with discreet irony, then, that in the Prologue Gower recalls an alternative classical precedent – not of the poet as an urbane observer, but as a chosen vessel of communication between the gods and man, epitomised in the career of the poetic archetype of Arion;

> Bot wolde god that now were on
> An other such as Arion,
> Which hadde an harpe of such temprure,
> And therto of so good mesure,
> He song, that he the bestes wilde
> Made of his note tame and milde,
> The Hinde in pes with the Leoun,
> The wolf in pes with the Moltoun,
> The Hare in pees stod with the Hound;
> And every man upon this ground

> Which Arion that time herde,
> Als wel the lord as the scheperde,
> He broghte hem alle in good acord,
> So that the comun with the lord,
> And lord with the comun also,
> He sette in love bothe tuo
> And putte awey malencolie.
> That was a lusti melodie,
> Whan every man with other low;
>
> (Prol. 1053–71)

The defining characteristic of Arion's performance is its apparent effortlessness, brooking no resistance in communicating its message of concord to the natural world and to every man, and it is that quality of effortlessness to which Gower felt unable, or perhaps unwilling, to aspire. Gower makes no claim to set himself up as a poet inspired by divine frenzy.[14] Indeed, in a moment of sly humour, his Muse intervenes in the poem only once, to suggest it would be better if he fell silent:

> My muse doth me forto wite,
> And seith it schal be for my beste
> Fro this day forth to take reste,
>
> (VIII 3140–2)

Though the image of Arion represents one persuasive model of poetic influence, Gower's purpose in briefly drawing it to the reader's attention within the *Confessio* is rather to enable him to define his alternative practice. In Gower's time, there is no 'other such as Arion', and the implication of the phrase is that such direct expression of poetic power lies in the past, perhaps in that golden age when

> The poeple stod in obeisance
> Under the rule of governance,
> And pes, which ryhtwisnesse keste,
> With charite tho stod in reste:
> Of mannes herte the corage
> Was schewed thanne in the visage;
>
> (Prol. 107–12)

and when not merely men, but even language, was not yet fallen:

> The word was lich to the conceite
> Withoute semblant of deceite:
> (Prol. 113–14)

Arion's example is one which Gower is barred from emulating, if not by his conception of his own poetic value, then by the passage of time, which has made such direct expression 'Withoute semblant of deceite' impossible.[15] Instead, and displaying the dogged diffidence which characterises his practice through the poem, Gower draws attention to the example of Arion before abandoning it as impractical for his needs. Where Arion's effect is achieved immediately and without apparent effort, Gower must construct for himself a validating myth more in tune with Christian notions of service, humility and effort. It is precisely this he achieves in the final meeting with Venus, in which his work is granted the approval of a semi-divine figure, or perhaps some aspect of divinity. Where Ovid's Arion seems almost in competition with Apollo, Gower's poetic persona proves far more respectful of the due proprieties of divine and literary hierarchy.

Compared to the triumphant command of Arion, Gower contrives to suggest an altogether different, more subtle ethical colouring to the resolution of his myth of authorship. Setting the poem's comprehensiveness against its consequent physical cost, which leaves the poet 'feble and impotent' (VIII 3127), the *Confessio* seems to exhibit some of the formal characteristics of a retraction – a partial denial of the value of literary fame. But this proves illusory, for the balancing of the natural man against the literary result is employed to precisely the contrary effect. In a charming reversal, Venus, the heathen goddess whose whim has created so many tragedies in the book, intervenes to advise the figure of the poet:

> Bot go ther vertu moral duelleth,
> Wher ben thi bokes, as men telleth,
> (VIII 2926–7)

Though she seems to dismiss him, in fact she is endorsing his life's work, in recognising the 'vertu' of the writings that have consumed his youth. As if in prophetic challenge to such (presumably unwritten) Canterbury Tales as those of the Clerk and the Man of Law, Gower asserts the value of his poetry through the testimony of Love herself that virtue is not simply a question of innate goodness but can be taught, as Aristotle sought to teach Alexander.[16] The apparent retraction is redeemed in the validation accorded his work, and the relationship of life and art is redefined as implicitly sacrificial – after all, a young man could never have amassed either such a great store of matter or the wisdom to display it to such positive effect.

In devising this complex ritual through which his work is granted value in proportion to the time and effort Gower has expended, the poet reveals the limitations of Foucault's remark quoted at the start of this essay. In a curious display which combines self-assertion and humility, Gower succeeds in creating a work of art not from the events of his life, but from the self-denial of action which enabled him to achieve a literary career. Whereas, in Chaucer's poetry, the idea of the narrator as being exiled from the action is constantly and humorously invoked,[17] Gower goes beyond this device to create a work in which the value of the work is explicitly related to that foregoing of life which has enabled its writing. The result is audacious, displaying the inevitably mixed nature that has characterised every element of the poem. In the vision of the poet swooning after being shown his true state in the mirror:

> Wherinne anon myn hertes yhe
> I caste, and sih my colour fade,
> Myn yhen dymme and al unglade,
> Mi chiekes thinne, and al my face
> With Elde I myhte se deface,
> So riveled and so wo besein,
> That ther was nothing full ne plein,
>   (VIII 2824–30)

Gower achieves, for a moment, an unaccustomed note of vulnerability, which must however be recognised as constituting one element in the poem's artful strategy to exploit the affective possibilities of the auto-biographical mode as a means to establish his poetic value.

For even in this final tableau of the poet and Venus, there is a further level of irony at work. Gower parodies his poetic equivalent as an old man and failed lover, but his recognition of old age is preceded by the appearance of a company of lovers:

> With him cam al the world at ones
> Of gentil folk that whilom were
> Lovers, I sih hem alle there
> . . .
> Ther was Tristram, which was believed

> With bele Ysolde, and Lancelot
> Stod with Gunnore, and Galahot
> . . .
> Wher Paris stod with faire Eleine,
> Which was his joie sovereigne;
> And Troilus stod with Criseide,
> . . .
> With hem I sih wher Dido was,
> Forsake which was with Enee;
> (VIII 2456–8, 2500–3, 2529–32, 2552–3)

the children of men's love for their art. There is something both surprising and moving in Gower's calling together of figures drawn from his tales – Medea, Deidamia, Narcissus, Canace – not freed from the consequences of their predicament, but granted a new extension of life through the power of Gower's reincarnating art. But the suitors at Venus' court are not merely the cast of the Confessor's tales. At the tailend of this demonstration of the alternative procreative power of Art appear the great literary authorities of the past – Aristotle, Virgil, Plato and Ovid – to plead his case:

> And whan thei comen to the place
> Wher Venus stod and I was falle,
> These olde men with o vois alle
> To Venus preiden for my sake.
> (VIII 2726–8)

These authors are strangely characterised as 'These olde men', which is of course what Venus' mirror has revealed Gower to be also, and through the implied comparison, Gower is gathered into the company of his esteemed mentors. In doing so, Gower presents an embodiment of the words with which the poem opens:

> Of hem that writen ous tofore
> The bokes duelle, and we therfore
> Ben tawht of that was write tho:
> (Prol. 1–3)

for the presence of the literary ancestors imagined in the poem's close lives on through the assimilation of their work into the books of later authors, a process witnessed in the text we have been engaged in reading.[18]

But in presenting himself as having reached a physical state comparable to theirs, Gower seeks to appropriate for himself the prestige that passing time has accorded their work. He had previously forged a connection between the linked achievement of literary eminence and advanced age in his curiously phrased[19] praise of Chaucer in early versions of the *Confessio*:

> And gret wel Chaucer whan you mete,
> As mi disciple and mi poete:
> ...
> For thi now in hise daies olde
> Thow schalt him telle this message,
> That he upon his latere age,
> To sette an ende of alle his werk,
> (VIII 2941–2*, 2950–3*)

but with the appearance of Virgil and Ovid in the congregation of *Elde*, the origins of this identification become clear. In the conclusion of a poem in which Gower has seemed to question the value of his own verse, through a final ironic twist he effects a vision of the enduring contemporaneity of literature – contrasted with the failing powers of the physical. What appears a depiction of physical 'truth' proves to be a statement of aspiration as Gower appropriates the classically derived prestige of the autobiographical mode to stake his claim to be classed with the immortals.

Notes

1   Michel Foucault, 'On the Genealogy of Ethics: A Dialogue', printed in *The Foucault Reader*, ed. Paul Rabinow (London: Peregrine, 1986) p. 370.

2   Headnote to Book 1 l.59. Translation from J.A. Burrow, 'The Portrayal of Amans in Confessio Amantis', printed in Gower's *Confessio Amantis: Responses and Reassessments*, ed. A.J. Minnis (Cambridge: D.S. Brewer, 1983) pp. 12–13.

3   In 'Lust and Lore in Gower and Chaucer', *Chaucer Review* 19 (1984) p. 110, Judith Davis Shaw, for example, reaches the judgment that 'Gower is so secure in his own vision that he never questions his own, not inconsequential role as creator.'

4   All quotations from the *Confessio Amantis* from *The Works of John Gower – The English Works*, ed. G.C. Macaulay, 2 vols (Oxford: Clarendon Press, 1901).

5   *Abelard's Ethics* – an edition with introduction, D.E. Luscombe (Oxford: Clarendon Press, 1971) pp. 32–3.

6 *The Complete Works of Geoffrey Chaucer,* ed. F.N. Robinson (Oxford: Oxford University Press, 1966), *The Canterbury Tales,* introduction to the Man of Law's Tale (B 88) p. 63.

7 C.S. Lewis, *The Allegory of Love* (Oxford: Clarendon Press, 1936) p. 201.

8 Rosemary Woolf, 'Moral Chaucer and Kindly Gower', from *Art and Doctrine: Essays in Medieval Literature,* ed. Heather O' Donoghue (London: Hambledon Press, 1986) pp. 197–218 (p. 202).

9 Yeager draws attention to the derivation of the title *Vox Clamantis* from the writings of the most revered of poetic prophets, Isaiah, extrapolating from this fact a characterisation of the tone of Gower's two previous extended poems: 'in all the poetry of Gower that Chaucer could have read by the time he completed the dedication, Gower's authorial attitude is outspokenly combative. He does not cajole or create allegories and then leave them for his readers' interpretation; this he does later, in the *Confessio.* In his earlier works, Gower uses the first person to declaim.' ('"O Moral Gower" – Chaucer's Dedication of *Troilus and Criseyde', Chaucer Review* 19 (1984) p. 87.)

10 Lewis, p. 201.

11 *The Poems of Robert Henryson,* ed. Denton Fox (Oxford: Clarendon Press, 1981), *The Testament of Cresseid,* pp. 344–50.

12 Lines from 'Love and Life', *The Complete Poems of John Wilmot, Earl of Rochester,* ed. David M. Vieth (New Haven: Yale University Press, 1968) p. 90.

13 For example, *The House of Fame* II 614–28 (Robinson, p. 288) and *The Parliament of Fowles* 4–11 (Robinson p. 310).

14 Medieval interpretations of classical notions of the divine madness of poets are considered by Ernst Robert Curtius, *European Literature and the Latin Middle Ages,* transl. Willard R. Trask (London: Routledge & Kegan Paul, 1953).

15 Yeager inventively describes the figure of Arion as a pointer, embedded early in the poem, to illuminate Gower's conception of the poet's role. In such a reading, Gower's call – 'Bot wolde god that now were on / An other such as Arion' – represents a prophecy which is seen to be fulfilled in the course of the poem, revealing 'at once Amans/Gower and *mutatis mutandis,* the poet John Gower himself'. Though the image of Arion is prominently placed, at the close of the Prologue, there is still something surprising in Yeager's insistence that this short episode provides the key to the aesthetic of the whole poem. There is surely a fundamental difference between the absolute capacity of Arion to inspire peace in men's hearts, and the more tentative, beseeching tone of Gower's plea for peace, with its appeal for divine grace:

> As he which is of alle thinges
> The creatour, and of the kynges
> Hath the fortunes uppon honde,
> His grace and mercy forto fonde
> Uppon my bare knes y preie,
> That he this lond in siker weie
> Wol sette uppon good governance.
> For if men takyn remembrance
> What it is live in unite,
> Ther ys no staat in his degree

> That noughte to desire pes,
> With outen which, it is ni les,
> To seche and loke in to the laste,
> Ther may no worldes joye laste.
>
> (VIII 2981–94)

R.F. Yeager, *John Gower's Poetic: The Search for a New Arion* (Cambridge: D.S. Brewer, 1990) p. 238.

16 An account of Aristotle's education of Alexander provides the material for Book VII of the *Confessio Amantis*.

17 For example, *The House of Fame* Book III, Prologue to *The Legend of Good Women*, *The Parliament of Fowles* ll. 5–10, *Troilus and Criseyde* Book III 1310–37.

18 The fruitful complexity of the relationship of the *Confessio* to anterior sources is strikingly recognised in J. Simpson's characterisation of the poem, in *Sciences and the Self in Medieval Poetry* (Cambridge: Cambridge University Press, 1995) p. 254, as an expression of a highly sophisticated reading strategy: 'What else is the *Confessio Amantis* than, at one level, an extended and extremely subtle account of the psychology of reading?'

19 Gower's apparent depiction of the pre-*Canterbury Tales* Chaucer as an old man contemplating the end of his endeavours drew indignant strictures from Elizabeth Barrett Browning in her unsigned review of *The Book of the Poets*, in *The Athenaeum* No. 762, 4 June 1842.

2
The Construction of an Author: Pietro Aretino and the Elizabethans

Harald Hendrix

Writing about oneself in an age when autobiography was at most a hybrid category means taking a stand as well as trying to define one's own identity. The two motivations often mingle, and they are inseparable in the case of people that perceive of their own identity as fundamentally different or distinct. Self-representation easily becomes apologetic and polemic. As such an act of defiance, we find the autobiographical urge in what is no doubt the most impressive text the Italian Renaissance produced in this (until then) informal genre, Cellini's *Vita*. Written around 1560 as a celebration of the extraordinary accomplishments of this Florentine goldsmith and sculptor, the text documents the unprecedented social advancement of the Renaissance artist as much as the psychological attitude of arrogant boasting which seems to come naturally with it. Cellini's shameless but highly amusing inclination to exaggerate his own achievements should, however, not only be seen as a reflection of his pathological conceitedness, but as an indication that his autobiographical writing springs from the urge to construct as well as document his own personality. In the exaggeration we may uncover his ambitions: what he is constructing by way of hyperbolic expression is not just an ordinary private person, but an artist worthy of praise and admiration, and especially of social and material recognition.

Although Cellini's *Vita* did not in any way inaugurate a tradition of artistic autobiography (it was published only as late as 1728), it reflects a mechanism of self-representation which can be detected all over Cinquecento Italy as well as in other parts of early modern Europe, including Elizabethan England. Closely related as it is to the social emancipation of artists (painters, sculptors and poets alike), it not only addresses issues of artistic innovation and excellence but also, and often quite emphatically, the relation between artist and patron. The pursuit of material and intellectual independence is actually one of the main components in this kind of idealised self-portrait, together with a demonstration of artistic skill and – in some rather more exceptional cases – a discussion of apparently random aspects of day-to-day life. The self-portrait as such is, of course, the best-known expression of this mechanism of self-representation, and Rembrandt painting himself in the pose of Titian's Ariosto – as in the London canvas – is indeed a very fine example. But perhaps even more revealing are some other ways of presenting oneself to the world: just think of the commanding dwellings of painters like Vasari in Arezzo and Florence or Rubens in Antwerp, and it will be clear that the private building activities of artists say less about their personal habits than about their social and artistic aspirations.

Early modern self-representation by artists is more about constructing oneself according to one's boldest ambitions than about revealing the intimate self. As such, it may well be regarded an innovation which during the sixteenth and early seventeenth centuries only gradually emerged, since in that period there were not yet any standard ways of expression. Cellini writes a book on his life, Rembrandt paints a series of self-portraits and Rubens builds himself a house; all are models of self-representation, private and public at the same time, that ultimately define one's artistic personality and project it in a more or less proud or provocative way onto the outer world. But although this was a new phenomenon with regard to the means of expression, it was not without models. Painters could recall the mythical example of Apelles and Alexander the Great – as Vasari actually did in a fresco on one of the walls of his home in Florence – and poets could follow in Martial's footsteps pleading a revival of the times of Virgil and Mecenas: 'Sint Maecenates, non derunt, Flacce, Varones.' In aligning themselves with their mythical or classical ancestors, painters and poets grasped the opportunity not only to boast of their talents, but to voice their hopes of social and material advancement as well.

Among the literary models available as touchstones in this early modern operation of artistic self-representation, the Italian author Pietro Aretino stands out not only for being a contemporary, but above all because he had been advocating a quite unorthodox strategy of self-promotion, based on a rather bold and aggressive manner of handling his patrons. Although well known all over Europe, his fame seems to have appealed in particular to early modern England, where from the late Tudor to the early Jacobean period there is a constant attention – positive and negative – to what Aretino represents.[1] Even though we have only rare evidence of straightforward imitations, there can be little doubt that the image he carefully constructed of himself, especially in his *Letters*, became a guideline for many an Elizabethan poet and intellectual who was struggling to define his own position. In discussing here first the appearances of Aretino's English reputation and subsequently some of the more characteristic aspects of his strategy of self-representation, I will not suggest that the example set by Aretino became a well-defined model widely imitated by the Elizabethans, but that in a more general way it set a standard which also in England appealed to many a man of letters concerned with his social and artistic identity.

This standard was effective and controversial at the same time, as is illustrated in the fierce polemic between Gabriel Harvey and Thomas Nashe in the 1590s, a quarrel that centred in many ways on the definition of the Elizabethan writer's artistic persona.[2] While at first only one among several names of model writers, in the ever-increasing vehemence of the debate Aretino gradually becomes the focus of controversy, representing for Harvey a 'monstrous wit' and for Nashe 'one of the wittiest knaves that God ever made'.[3] This famous definition from *The Unfortunate Traveller* earned Nashe the reputation of being 'the true English Aretine',[4] an epithet he no doubt welcomed as a stimulus in his striving for independence, both intellectual and material; that, at least, seems to be the essence of his identification with Aretino, as can be deduced from one of his first comments on the Italian author:

> We want an Aretine here among us, that might strip these golden asses out of their gray trappings, and after he had ridden them to death with railing, leane them on the dunghill for carion. But I will write to his ghost by my carrier, and I hope hele repaire his whip, and use it against our English Peacockes, that painting themselves with church spoils, like mightie mens sepulchers, have nothing but Atheisme, Schisme, hypocrisie, and vain glory, like rotten bones lie lurking within them.[5]

Nashe's praise and admiration are counterbalanced by Harvey's biting remarks both on Aretino and on his English epigone, almost seen as one and the same person:

> Cannot an Italian ribald, vomit-out the infectious poyson of the world, but an Inglishe horrel-lorrel must licke it up for a restorative; and attempt to putrify gentle mindes, with the vilest impostumes of lewde corruption?[6]

Harvey's unusually harsh tone suggests not only that he is about to yield to his opponent, as in fact is the case, but also that the thought of Aretino was able to awaken fierce emotions among the later Elizabethans. Less clear, however, is what peculiarities of the Italian writers *oeuvre* were provoking such reactions, and what someone like Harvey was referring to when writing about 'the infectious poyson of the world' or 'the vilest impostumes of lewde corruption'. Addressing this question becomes all the more appropriate when we see that some years before his quarrel with Nashe, Harvey expressed quite different judgments on Aretino, celebrating him, for example, for his 'singular extraordinarie veine and invention', opinions that reflect, it has been suggested, a more general attitude of admiration towards the Italian.[7]

The inconsistency we can detect in Harvey's judgments on Aretino is by no means an isolated phenomenon, as the case of Ben Jonson should remind us. The overt condemnation of the Italian as a repulsive pornographer in *Volpone*[8] did not prevent Jonson from extensively consulting Aretino's theatrical production while preparing his *Epicoene*, adopting not only precise motifs like the central one of a cross-dressed boy serving as a bride, but also more general qualities pertaining to the dramatic structure.[9] In all its ambiguity, this attitude towards Aretino reflects the overall reaction of the later Elizabethans to Italian culture.[10] After the publication of Roger Ascham's *The Schoolmaster* in 1570, anti-Italian sentiment became increasingly fashionable, partly as a result of growing anti-Catholicism, partly as a consequence of a number of book publications that shed a negative light on some of the once venerated heroes of the Italian Renaissance. Ascham's book itself gave another and a much less favourable meaning to the concept of 'courtier' as first propagated by Castiglione's canonical dialogue;[11] by fiercely attacking Machiavelli's *Principe*, Gentillet's *Contre-Machiavel*, immediately translated into English in 1577, discredited Italian politics no less than the concept of 'politic' as such.[12] Finally, John Wolfe's 1584 publication of Aretino's *Ragionamenti* no doubt contributed considerably, albeit unintentionally, to propagating the author's fame as a pornographer, a reputation so notorious that even Wolfe's other Aretino editorial enterprise, the celebrated 1588 publication of four of his plays, was unable to correct it any more.[13]

The loud and general condemnation of Aretino by late Elizabethan and early Jacobean culture is, however, but one part of a more complex situation, as is convincingly demonstrated by the unmistakable impact Wolfe's edition of Aretino's plays had on contemporary English theatre.[14] Ever more eager to develop an identity of their own, the later

Elizabethans welcomed the possibility to look upon foreign models condescendingly, although using them at the same time for their own purposes. Aretino was widely loathed for his erotic production, but his theatre was admired and imitated, although only a few had the courage – or the modesty – to admit that. There were, however, more elements to what one could call the 'model' Aretino. The fame and impact of his theatre and pornographic dialogues started in England with the two publications by John Wolfe in 1584 and 1588, but Aretino's name was already familiar to many an Englishman well before that time. As an artist he was hugely respected and envied, though not for his plays or his *Ragionamenti*, but for his peculiar style and originality, for his acuteness and his independence.

This is apparent even in the early reactions of Gabriel Harvey, those preceding his polemic with Nashe and dating from about 1580. He appreciates Aretino as one of 'the most delicate, and fine conceited Grecians and Italians' on account of his (as already quoted) 'singular extraordinarie veine and invention', or as an author 'singular for rare and hyperbolical amplifications'.[15] This reference to 'hyperbolical amplifications' makes it easy to relate Harvey's praise to a group of prose narratives on religious subjects renowned for their particularly emphatic style. The publication of these texts, between 1534 and 1543, not only brought Aretino great fame and some fortune, but launched many efforts to imitate and emulate them. Among their admirers was the poet, courtier and diplomat Sir Thomas Wyatt, who in the late 1530s composed his *Penitential Psalms* in part as a translation in verse and in part as a continuation of Aretino's *I sette salmi de la penitentia di David* (1534), the result being published posthumously in 1542. Wyatt's decision to model his version of the psalms on the paraphrase published by Aretino just a few years earlier suggests that he too admired the unorthodox style Aretino had introduced in that work, and that this appraisal was not uncommon in circles of the late Tudor court.[16]

Indeed, there can be little doubt that Aretino was highly regarded by leading Tudor courtiers including Thomas Cromwell, Philip Hoby and William Paget, and possibly even by Henry VIII himself.[17] We can only speculate on their motives for supporting him, but we are well informed on Aretino's enthusiasm for his English admirers. Keen as he always was on easy money, he initially hoped that after the suppression of the monastic orders in 1539 Henry would grant him a sinecure, and once these hopes had been disappointed, he tried another strategy and in 1542 dedicated the second book of his *Letters* to the English monarch. Henry subsequently expressed his gratitude by awarding him – four years later – the considerable sum of £75, which as a result of various delays arrived in Venice only at the end of 1547, and thus after Henry's death.[18] This generous gift, donated as it may have been out of sincere admiration for the author's talents or as an attempt to make him into one of the King's clients, contributed considerably to Aretino's fame in England, not as a token of literary excellence, but as a demonstration of the artist's independence from patrons. In a letter dedicated to Aretino and published by way of introduction to his *The Pilgrim*, the well-informed William Thomas, author of a *Historye of Italy* (1549) and of the *Principal Rules of the Italian Grammar* (1550), refers to this episode when stating:

> because I understand that the King, in defense of whose honour I have made it [*The Pilgrim*], hath remembered thee with an honourable legacy by his Testament; the which his enemies pretend proceeded from the fear that he had lest thou shouldest, after his death, defame him with thy wonted ill speech.[19]

By denouncing the rumours of a possible blackmail,[20] Thomas referred to Aretino's fame as an astute extortioner who by appealing to public opinion was able, with his *Letters*, to force patrons to pay him. This idea of an artist who exploited ill speech to his own benefit dominated Aretino's early Italian reception, and the reference in Thomas's dedication suggests that it also affected his English reputation. Such a conjecture is confirmed by later comments such as Nashe's judgment that Aretino's 'pen was sharp-pointed like a poniard; no leaf he wrote on but was like a burning-glass to set on fire all his readers',[21] and one might even be tempted to read Gabriel Harvey's outburst on the 'infectious poyson of the world' as an indication that even when the Italian's reception was completely dominated by allegations of pornography, the 'ill-speech' argument was still present.

There is, moreover, another significant correspondence between the reactions of Thomas, Nashe and Harvey: they all praise Aretino's 'naturalness'. In his dedication letter, written about 1550, Thomas compliments Aretino by saying: 'thee, whose virtue consisteth only in Nature without any art, than unto any other; whom I know both natural, virtuous, and learned withal.'[22] The compliment is repeated by Harvey in 1574, commending 'Aretine's glory to be himself: to speak and write like himself: to imitate non but himself and ever to maintain his own singularity; yet ever with commendation and compassion of others.'[23] Finally, in the famous phrases from *The Unfortunate Traveller* (1594), Nashe remarks with astonishment: If out of so bas a thing as inke there may be extracted a spirite, hee writ with nought but the spirit of inke; and his stile was the spiritualitie of artes, and nothing else...'.[24]

Aretino was thus admired for his originality: he was perceived as an author who did not lean on tradition, but used only his personal talents. This image of a 'natural' poet appealed to the Elizabethan men of letters as much as to their Italian colleagues. But it was a false image. Aretino certainly did not only work with his talents. He was, on the contrary, one of the first men of letters to put into practice a kind of industrial writing, using not only many literary sources but even a team of ghost-writers.[25] The image was nevertheless strong, constructed and propagated as it was by Aretino himself, and the efficiency of his strategy is convincingly demonstrated by Harvey's just quoted statement 'to imitate non but himself', repeating the exact words Aretino himself had used in one of his *Letters*.[26]

The early modern English reaction to Aretino turns out to be much more complex and certainly more positive than the caricature-like condemnation of the author as a 'poysonous ribald' mainly renowned for being a pornographer would suggest.[27] The stylistic innovations introduced in his religious narratives gained Aretino wide acclaim, from the 1540s to the late Elizabethan era, when accusations of obscenity tended to overshadow the earlier appraisal. But as the cases of Gabriel Harvey and Ben Jonson demonstrate, there is sufficient reason to believe that even after 1585 the fierce criticism was but a superficial reflection of the more general anti-Italian mood, and could not wipe out the fascination with the novelty Aretino represented. Apart from the particular stylistics of his religious production, this quality was identified with his theatre, which actually provoked various kinds of imitation, and with his strong image as an original, 'natural' writer who had been able to achieve, not with flattery but only with his own genius, financial and intellectual independence.

This image, which had a particular appeal – so it seems – to 'homines novi' like William Thomas and Thomas Nashe, was the result of a conscious strategy of self-promotion by a man who himself was perhaps the epitome of the Renaissance 'new man of letters'. Born in 1492, Aretino did not belong to the generation of those writers who, like Ariosto and Castiglione (born in the 1470s) established many of the conventions of Renaissance literature and culture in general. This perhaps explains why his position was quite unconventional from the start of his career. His ambition to establish himself in Rome as a leading court poet was soon disappointed, not only by the election of the austere Dutch pope Adrian VI in 1522, but also by growing opposition to his quite disrespectful and malignant satirical production, the so-called 'pasquinate'. This eventually forced him to leave Rome and the papal court in 1525 and seek protection, first in the intimate friendship with the leading warlord Giovanni de' Medici, then, after the dramatic death of his friend, at the court of the Gonzagas in Mantua, and finally, in 1527, in the free republic of Venice, where he would stay for almost 30 years till his death in 1556. These circumstances had a great influence on Aretino's literary production. Whereas in the ten years between 1517 and 1527 he tried to make a living as a court poet, producing texts appropriate to this specific cultural environment, in his Venice years he gradually evolved into something of a freelance writer – a completely new phenomenon – who was not entirely dependent on material support from patrons, but was also trying to make good money from the sales of his books.

This switch in economic orientation explains the hugely experimental drive we can detect in Aretino's production during the 1530s. He continued to write theatre for the carnival season in various Italian courts, while starting to experiment not only with religious, but also pornographic, literature in exactly the same years. The first part of his *Ragionamenti* was published in 1536, but already two years earlier, in 1534, the first of his religious narratives – on the Psalms and on the Life of Christ – were on the market. There can be little doubt that Aretino, with this quite diverse production, was trying to address different audiences, in order to extend the range of potential buyers of his books. The considerable and in the long run most influential literary innovations he accomplished in this way were thus primarily dictated by economical considerations and not by discontent with the literary status quo, although this certainly did contribute to the choice of the specific alternative forms he adopted.

Perhaps the most interesting part of Aretino's experimental production is his *Letters*. From the moment when, in 1525, he fled the hostile papal court in Rome, Aretino started an intense correspondence with those whom he was forced to leave, and increasingly with those from whom he hoped to gain material or other support. This correspondence remained informal for some ten years, until around 1536 the idea to publish his private letters came to mind as one of the possible strategies for inventing new and commercially attractive editorial enterprises. This was indeed a brilliant idea. The first volume of his *Letters*, which came out in 1538 and contains 344 letters dating from 1525 up to that same year, was an overwhelming success. It was to be followed by a second volume in 1542, the one dedicated to Henry VIII, just as successful as the first one, and there were still four other volumes to follow, the last being published posthumously in 1557. So we have a total of six volumes, with almost 2,000 letters over a period of some 30 years.

The most striking and indeed distinctive feature of Aretino's *Letters* is their commercial exploitation. Most of the letters Aretino wrote after 1536, when he embraced this project, were written with the idea of immediate publication and had therefore a twofold destination: the person to whom the letter was addressed, and the general public who bought the published version. This of course explains why these texts must be regarded as consciously premeditated forms of self-representation. But the same is true of the earliest letters, those dating from before 1536, when the project to publish them took shape. And this is not only because Aretino, when preparing the edition of the first volume, took considerable liberty in altering his original letters. It is also because, precisely in these first letters of a still more personal nature, he had already accomplished a way of presenting himself – be it by chance, by talent or by premeditation – which was not only highly regarded by his correspondents, but also gave him the basic components of the formula which he was to exploit to the full later on, with the publication of the six volumes. Among these components perhaps the most prominent are, first, the emphasis on his own extraordinary sensibility and on his 'natural' literary talents and, second, the suggestion that he himself is a perfect example of a universal personality whose companionship is highly valued not only by kings and princes but also by intimate friends and by his family.

These particular aspects are to a large extent responsible for the reputation Aretino has enjoyed since Burckhardt, who celebrated him as archetypical modern man, with his libertarian morals, his socially ambitious attitude and his unusually fine observation of nature and human character. And there is indeed much to be said for such a judgment. Whoever reads, in the very first part of the first volume of his *Letters*, the famous account of the dramatic death of Aretino's friend and protector, Giovanni de' Medici, cannot but be impressed by the highly compelling manner in which the author renders this tragic episode and by the deep grief he is able to communicate to his correspondent.[28] And the equally famous letter to Titian of some years later (1544), in which Aretino describes the magnificent view from his Venetian house on the Grand Canal, demonstrates such an unusual attention to the charms of city life, that it has been rightly pinpointed as an absolute starting point of modern sensibility for the beauties of ordinary urban life.[29]

There can be little doubt that Aretino was an author who in many ways anticipated or even generated literary developments which would only many years later, and in some aspects even centuries later, find general approval and imitation. Though surely the product of an extraordinary talent, it is clear nevertheless that these brilliant innovations could materialise because of the economic pressure the author was exposed to. He had to make money as an independent man of letters, and could do so not only by writing as much as possible, but also – and that was a revolutionary insight – by constructing a particular image of himself, by presenting himself as a brilliant writer whom no one, not even the emperor or the pope, could refuse material and moral support. By posing as a good author, he became one in the eyes of his correspondents, and was able to persuade them to support him. He effectively constructed his own image in order to capitalise on it, and the success of this strategy largely contributed to his reputation, in Italy as well as in other parts of sixteenth-century Europe, such as Elizabethan England.

But how exactly did Aretino construct his self-image as an author? And why was it so successful? In my discussion of this point I will concentrate on the two strategies mentioned earlier on: his ability to convince his readers of his exceptional but 'natural' literary talents, and his inclination to project himself as a complete personality. This last point is of central importance in the many letters where Aretino discusses, in a seemingly informal manner, all kinds of matters relating to his day-to-day life. Though mostly of an economical nature, these discussions present a fascinating mixture of affections and financial preoccupations where he writes about his two daughters, Adria and Austria, born in 1537 and 1547 from a protracted love affair with his already married maid, Caterina Sandella. Aretino not only expresses great parental love for his offspring, but also shows himself extremely concerned to guarantee his bastard daughters financial security, writing many letters to patrons, mostly women, in order to win them a substantial dowry, and not without success, for he gained for each the impressive sum of 1,000 scudi. What is striking in these letters is the presence of many everyday details which at first glance do not seem relevant to the purpose of the text. But the fact that these letters were hugely successful – they brought Aretino the money he wanted – suggests that these seemingly irrelevant details were perhaps part of a rhetorical strategy. They underscore the fact that Aretino is a very caring parent, not just by saying so, but by showing it in practice. This is a strategy that worked well with his patrons, the direct addressees of the letters, but also with the general public. This is well illustrated in a letter that seems quite exceptional, precisely because it is such a normal letter. It was written in December 1547 to Caterina Sandella, and discusses matters of such a normal, everyday nature that it is quite a surprise to find it published next to letters written to the emperor and the pope.

> I, Caterina, pray to you that you do not pray me, and enjoin you that you do not enjoin me, and counsel you that you do not counsel me to give Austria to the nurse we used to have and Adria to the school-mistress we have now. The former is dry of milk and the latter is a wench in her habits. Let my two daughters be brought up in our own house. Then the younger will be fed upon endearments and the elder will be taught with love.[30]

This letter, obviously, is more complex than it might seem at first glance. Aretino discusses with the mother of his children quite private matters relating to their education, things one might not expect to be published in a collection of letters sold to the general public. This indic-ates the author's ambition to be seen as a loving father, an image he most probably considered to be appealing to readers and therefore a commercially attractive kind of self-representation.

But there is more. The letter presents quite conspicuous stylistic fea-tures. The first sentence consists of a sequence of paradoxes, a rather emphatic way of addressing a person, and a formula most husbands would be hesitant to use when writing to their wives. Here Aretino gives proof of his brilliance as a writer, even in a private context, which might be taken as a demonstration of the other rhetorical strategy mentioned before: Aretino's technique to present himself as an excep-tionally talented 'natural' writer. This is a strategy he uses mainly to impress (possible) patrons, but he certainly does not refrain from using it also in other contexts in order to baffle the general audience. The essential component of this technique is paradox. By using in his letters complicated, even arduous, but always elegant paradoxes, Aretino not only amuses his addressees but gives at the same time tangible proof of his own literary versatility, thus convincing his patrons that their money is indeed well spent. This is a feature which dominates much of his correspondence, from the earliest days on, and was thus – as a tech-nique – not invented for the editorial exploitation of the letters, but for the specific goal of addressing and indeed dazzling his patrons. At times Aretino does not hesitate to write in a quite aggressive and almost insulting manner to those from whom he hopes to get money. At first glance this might seem not very wise, but it was after all extremely effective, since it did produce a constant flow of financial support. One of his earliest patrons, Guido Rangone, an aristocrat from Modena, continued to support Aretino, even after receiving thank-you notes like this one written in 1529:

> Since there is more happiness in giving than in receiving, I sincerely hope that from this gift that you sent me of money, of a coat of arms and of a frock of white satin, might arise in you the heighest grade of consolation. And it is your great chance that such is the virtue of courtesy; because when you apply yourself in the generosity of continually giving, you will continually be happy. And that is why I would give insult to your Lordship in renewing my expressions of gratitude for something which, since I have agreed to accept your gifts, should cause you to thank me.[31]

In writing down such elaborate and apparently insulting paradoxes, Aretino was giving the ultimate demonstration of his rhetorical skills. His patrons loved him for it, and they accepted without hesitation the manipulated self-representation which Aretino had so brilliantly concocted. Other early modern men of letters looked upon his example with envy and astonishment: 'Princes hee spard not, that in the least point transgrest,' as Thomas Nashe put it.[32] Aretino was clever indeed, and his example should not only be taken into account when examining the manner in which Elizabethan and Jacobean men of letters tried to define their own social and intellectual position, but should also once again remind us that, if early modern autobiography is a genre at all, its 'natural' self-representation is more often highly meditated and wittily constructed.

Notes

1   Edward Meyer, *Machiavelli and the Elizabethan Drama* (Weimar, 1897) p. xi, claims to have collected over 500 references to Aretino, against only 395 references to Machiavelli. A survey of the English, particularly the Elizabethan, reception of Aretino is still lacking. For a discussion of aspects of this reception, see Claudia Corradini Ruggiero, 'La fama dell'Aretino in Inghilterra e alcuni suoi influssi su Shakespeare', *Rivista di letterature moderne e comparate*, 29 (1976) pp. 182–203; Maria Palermo Concolato, 'Aretino nella letteratura inglese del Cinquecento', in *Pietro Aretino nel cinquecentenario della nascita* (Rome, 1995), vol. I, pp. 471–8.
2   For a reconstruction of this episode, see David C. McPherson, 'Aretino and the Harvey-Nashe Quarrel', *PMLA*, 84 (1969) pp. 1551–8.
3   Gabriel Harvey, *A New Letter of Notable Contents* (1593), in *Works*, ed. A.B. Grosart (London, 1884), vol. I, pp. 272–3; Thomas Nashe, *The Unfortunate Traveller and Other Works*, ed. J.B. Steane (London, 1971) p. 309.

4   A comment by Thomas Lodge in his *Wits Miserie* of 1596, two years after publication of Nashe's *The Unfortunate Traveller* (1594); see McPherson, p. 1555.

5   Thomas Nashe, *Pierce Pennilesse his Supplication to the Devill* (1592), in *Works*, ed. R.B. McKerrow (London, 1904), vol. I, p. 242.

6   Gabriel Harvey, *Works*, ed. A.B. Grosart (London, 1884), vol. II, p. 91.

7   Gabriel Harvey, *Works* vol. I, p. 93; *Marginalia*, ed. C.G. Moore Smith (Stratford, 1913) p. 124. For the evolution in Harvey's attitude toward Aretino, see in particular McPherson, 'Aretino and the Harvey–Nashe Quarrel'.

8   'some young French-man, or hot Tuscane bloud, / That had read Aretine, conn'd all his printes, / Knew euery quirke within lusts laborinth, / And were profest critique, in lechery'; Ben Jonson, *Works*, ed. C.H. Herford and Percy Simpson (Oxford, 1937), vol. V, III.vii. 59–62. See Mario Praz, *Machiavelli in Inghilterra* (Florence, 1962) pp. 195–210 ('L'Italia di Ben Jonson') especially pp. 207–9; M. Bervellier, 'Influencias italianas en las comedias de Ben Jonson', *Filosofia y Letras*, 3 (1942) pp. 5–72.

9   Oscar James Campbell, 'The Relation of *Epicoene* to Aretino's *Il Marescalco*', *PMLA*, XLVI (1931) pp. 752–62; Donald A. Beecher, 'Aretino's Minimalist Art Goes to England', in *Pietro Aretino nel cinquecentenario della nascita* (Rome, 1995), vol. II, pp. 775–85; Christopher Cairns, 'Aretino's Comedies and the Italian "Erasmian" Connection in Shakespeare and Jonson', in *Theatre of the English and Italian Renaissance*, ed. J.R. Mulryne and Margaret Shewring (London, 1991) pp. 113–37.

10 George B. Parks, 'The First Italianate Englishmen', *Studies in the Renaissance*, 8 (1961) pp. 197–216; Leo Salingar, 'Elizabethan Dramatists and Italy: A Postscript', in *Theatre of the English and Italian Renaissance*, ed. J.R. Mulryne and Margaret Shewring (London, 1991) pp. 221–37, especially p. 228.

11 Peter Burke, *The Fortunes of the Courtier* (Cambridge, 1995) especially pp. 110–16.

12 Praz, *Machiavelli in Inghilterra*, pp. 97–152 ('Machiavelli e l'epoca elisabettiana') especially p. 100: Gentillet's translation by Simon Patericke, though produced in 1577, was published only in 1602; the text of the *Contre-Machiavel* was nevertheless available in the Latin translation.

13 For information on this edition, see Pietro Aretino, *Sei giornate*, ed. Giovanni Aquilecchia (Bari, 1969) pp. 393–401. On its impact, see S. El-Gabalawy, 'Aretino's Pornography and English Renaissance Satire', *Humanities Association Review*, 28 (1977) pp. 9–19.

14 See J.M. Lothian, 'Shakespeare's Knowledge of Aretino's Plays', *MLA*, 25 (1930) pp. 415–24; Marvin T. Herrick, *Italian Comedy in the Renaissance* (Urbana, Il, 1960) especially p. 227; Leo G. Salingar, *Shakespeare and the Tradition of Comedy* (Cambridge, 1974), passim.

15 Gabriel Harvey, *Works*, ed. A.B. Grosart (London, 1884), vol. I, p. 93; *Marginalia*, ed. C.G. Moore Smith (Stratford, 1913) p. 124.

16 On Wyatt's adaptation, see C. Marengo Vaglio, 'Originalità e tradizione nella poesia di Sir Thomas Wyatt', *Atti della Accademia delle Scienze di Torino*, 113 (1979) pp. 91–120; Rivkah Zim, *English Metrical Psalms. Poetry as Praise and Prayer, 1535–1601* (Cambridge, 1987) pp. 43–79.

17 Testimony to these contacts are the references in Aretino's correspondence, as recorded, for example, in the anthology Pietro Aretino, *Lettere*, ed. Paolo Procaccioli, 2 vols (Milan, 1991), *ad indicem*.

18   The gift is documented by a registration in the *Book of payments* made 14 November 1546: 'Sir Ph. Hobbie, to be paid to divers Spaniards in reward at their return into their country, with £75 to Petro Aretino who dedicate a book to the king's said Majesty'; see *Letters and Papers, Foreign and Domestic, of the Reign of Henry VIII*, XXI, 2 (London, 1910) p. 448 [f. 94].

19   [William Thomas], *The Pilgrim. A Dialogue of the Life and Actions of King Henry the Eighth*, ed. J.A. Froude (London, 1861) pp. 1–2.

20   For a reconstruction of this episode, see GianMaria Mazzuchelli, *La vita di Pietro Aretino* (Milan, 1830) pp. 67–9.

21   Thomas Nashe, *The Unfortunate Traveller and Other Works*, ed. J.B. Steane (London, 1971) p. 309.

22   [William Thomas], *The Pilgrim*, pp. 1–2.

23   Gabriel Harvey, *Marginalia*, ed. C.G. Moore Smith (Stratford, 1913) p. 156.

24   Thomas Nashe, *The Unfortunate Traveller and Other Works*, ed. J.B. Steane (London, 1971) p. 309.

25   On Aretino's writing techniques, see Christopher Cairns, *Pietro Aretino and the Republic of Venice* (Florence, 1985).

26   'Ed è certo che io imito me stesso.'

27   Another qualification by Harvey; *Works*, ed. A.B. Grosart (London, 1884), vol. II, p. 94.

28   The letter to Francesco degli Albizi, dated Mantua 10 December 1526, now in Pietro Aretino, *Lettere. Libro primo*, ed. Francesco Erspamer (Parma, 1995) pp. 16–27.

29   The letter to Titian, dated May 1544, in Pietro Aretino, *Lettere*, ed. Paolo Procaccioli (Milan, 1991), vol. I, pp. 531–3; see on the same subject also the letter to his landlord Domenico Bollani, dated 27 October 1537, now in Pietro Aretino, *Lettere. Libro primo*, ed. Francesco Erspamer (Parma, 1995) pp. 440–5.

30   Slightly altered translation by Thomas Chubb in Pietro Aretino, *The Letters*, ed. Thomas Caldecot Chubb (Yale, 1967) p. 238. The original: 'Io, Catarina, ti prego che non mi preghi, e ti impongo che non mi imponga, e ti consiglio che non mi consigli a dare Austria a la balia di prima, né Adria a la maestra di adesso: però che l'una è di latte sterile, e l'altra di costumi mendica. Sì che stiansi le due mie figliuole in casa nostra, che al manco quella sarà nutricata di vezzi, e questa disciplinata d'amore', now in Pietro Aretino, *Lettere*, ed. Paolo Procaccioli (Milan, 1991), vol. II, pp. 446–7.

31   The original: 'Essendo maggior la felicità del donare che quella del ricevere, io ho caro fuor di modo che dal presente degli scudi, de la impresa e del saio del raso bianco, che mi fate, nasca in voi il sommo grado de la consolazione. Ed è vostra gran ventura che tanto possa la vertù de la cortesia; perché facendo voi l'essercizio de la liberalità nel donar continuo, continuamente sete felice. Per la qual cosa farei ingiuria a la signoria vostra prolungandomi in ringraziarla di quello che, per aver accettato is suoi doni, merito di essere ringraziato io'; now in Pietro Aretino, *Lettere. Libro primo*, ed. Francesco Erspamer (Parma, 1995) pp. 49–50.

32   Thomas Nashe, *Works*, ed. R.B. McKerrow (London, 1904), vol. II, p. 265.

3
The Vocacyon of Johan Bale: Protestant Rhetoric and the Self

Peter Happé

John Bale's *Vocacyon* was written in a great hurry at a critical point in his life.[1] Its place in this volume may be justified because it is a personal view of his own experience over a period of just about a year. It is in no sense a full account of his life, or a measured reflection upon an eventful past, since its main objective is to bring out a number of key points of Protestant belief, and it is focused upon the risks and exigencies of the present. Nor does it set out to be complete or comprehensive of all the things which happened to him in this eventful year. A polemical purpose underlies most of what is found in the work, which is why I have put the word 'rhetoric' in the title. The book is an attempt to declare a faith and to influence others. It would be ingenuous, however, to suppose that Bale confines himself to the fate of the individual Christian, for there are many aspects of what he recounts and of the way of presenting them which have a broader politico-religious dimension or motive. On the grounds of the way his self is constructed and in relation to the political milieu, we might indeed accept that Bale's short autobiography is a 'cultural product'.[2] Indeed, this self is developed so as to describe a situation and influence it.

There is, however, the paradox that in spite of a vigorous attempt to convey personal experience and to show much detail about certain aspects of it, the work shares with many other autobiographies – perhaps in a way to suggest that this is inherent in the genre – a sense that the doings of the real-life John Bale are viewed from a distance. There is an eye to commentary and to making the most of what can be drawn from the experiences here assembled. This constitutes a distancing which is the best clue to the idea of 'self'. At the same time the commentary helps to create the self. And I want to suggest that the idea of

45

the self and the use of rhetoric are intimately linked. The principal link, I hope to show, is fiction, a word which I use with the primary sense of 'making'. On a theoretical note I have been much interested by Barrett Mandel's contention that autobiography is not fiction largely because readers always know when a work is fiction and when it is autobiography.[3] My feeling about Bale's book is that we know it is meant to be an autobiographical account, and yet the sense of its being arranged in such a determined manner arouses curiosity about how much is 'made up'. There is no reason, however, to suppose that curiosity about this might obscure the didactic impact.

We naturally find here particular characteristics of the Protestant ethos with its special emphasis upon the personal responsibility of the individual. Much of Bale's religious belief and teaching is based specifically upon the point that the individual Christian has direct contact with God's grace. In this view the function of the clergy was not to be intermediaries, but to stimulate direct contact and to encourage in a typically Protestant way the use of the Scriptures as a means of bringing this about. The sense of the self in the Protestant ethos may be further sharpened by the rejection of auricular confession, and the consciousness of persecution which intensifies the idea of a personal struggle. It is one which is seen as a means of spiritual development, by which God tests the individual soul.

Nevertheless the rejection of intermediaries may not be quite what it seems because there is a tension in *Vocacyon* between this notion and Bale's emphatic concept of the role of the bishop. He was sent to Ireland to be Bishop of Ossory, and it is a major preoccupation of his account to bring out the qualities to which he thought Protestant bishops should aspire, and to denigrate the failure of the unrepentant Catholic bishops he encountered during his stay in Ireland. He measures his own achievement against these objectives, and there is circumstantial detail which is designed to further them. This preoccupation has to take its place with a number of other polemical objectives, which I hope to explore.

Contexts

After a period of exile following the fall of Thomas Cromwell in 1540, Bale returned to England hoping for preferment shortly after the accession of Edward VI in 1547. But his earlier alienation of Sir William Paget, Principal Secretary, over the martyrdom of Anne Askew in July 1546 may have delayed promotion until after the fall of Somerset when he was supported by John Ponet, now Bishop of Winchester.[4] In accordance with the new regime's policy of putting Protestants into positions of ecclesiastical authority, Bale was appointed in August 1552 to Ossory in southern Ireland. It was not an attractive appointment, and Bale claims in the *Vocacyon* that he tried to avoid it on the grounds of age and infirmity. His resistance as he reports it suggests a polemical point in showing that he was called to his new role by the highest authority, and not necessarily on the basis of his own qualities and merits. The royal imperative proving immutable, Bale set out to an environment which could hardly have been more inhospitable to his particular brand of Protestantism with its rejection of ceremony and his resistance to clerical celibacy.

If this 'vocation' to a bishopric is the first part of his autobiographical narrative, the death of King Edward in 1553 precipitated a new crisis which led to the critical events in the second part. This comprises the tightening of theological objections to his teachings by the Irish clergy, and the development, as he saw it, of conspiracy against him, followed by his flight, originally aiming for Scotland. As fate would have it he was seized by a rogue who took him pirating for several weeks in the English Channel until, by the deliverance of God, which Bale earnestly and triumphantly celebrates, he was released in the Netherlands on payment of a substantial ransom. He remained in exile for five years until the return of Elizabeth again changed the political climate.

The context of the *Vocacyon* in Bale's prolific writings must have affected the rhetoric in which it is expressed. By 1552 when the events started, he was 56 years old. Though an orthodox Carmelite friar until about 1534 he became, after his conversion, a part of Cromwell's propaganda campaign. In furtherance of the rise of Protantism in the aftermath of the royal divorce he wrote a large number of polemical plays, of which five have survived. During his first exile (1540–7/8) he turned his enormous intellectual energy to supporting the Protestant cause by a series of pamphlets vilifying in particular the conservative bishops such as Gardiner and Bonner, and advocating the continuing development of the Protestant state which more and more seemed threatened by Henry's own later conservatism. In support of this Bale reconceived British history to demonstrate the pollution of the true and original English Church by the missionary intrusion of St Augustine and later medieval Catholicism, and he set about continuing the work of John Leland, the King's Antiquary, by composing a catalogue of British writers many of whose works were now at risk following the break-up of monastic libraries at the Dissolution. This may have had a key role in the development of the *Vocacyon* which extensively refers to the historical structure Bale had formulated. His rejection of Catholicism turned largely upon his sense of historic errors.

When he returned to England in 1547 or 1548 after the death of Henry VIII, he showed some signs of wanting to return to his plays, but in fact he was still preoccupied with the need to extend the Reformation and rapidly became embroiled with some of the Hampshire clergy. Bale was overtaken by the public events I have noted, and there is a strong sense that he was affected by events as much as he was able to control them. His experience as a polemicist and his interest in a Protestant pattern of history were part of his stance as the author of the autobiographical account of his adventures in Ireland and the subsequent sea voyage.

Rhetoric

There were a number of models upon which Bale may have drawn for his autobiography. His work on Anne Askew and Sir John Oldcastle in the 1540s had sought to vindicate two figures from Catholic persecution by presenting them as 'saints' of the Protestant cause.[5] But for Bale the traditional role of saints as perpetrators of miracles had to be repudiated, and he was bitterly opposed to the worshipping of saints in order to prompt their intervention: this is why at one point he condemns de Voragine's *Legenda Aurea* in *Vocacyon*. Yet *exempla* there must be for the clarification of the new religion and this, I think, is why he adopts St Paul as a key model for his own affairs in the *Vocacyon*. The attraction of St Paul was, first, that the doctrine of justification by faith could be attributed to him and he had consequently been given high status by Luther and other leading reformers, and, second, that in his account of his journey from Jerusalem to Rome there could be found distinct parallels with Bale's own journey from Dublin to Flanders in 1553.[6] This is supported by rhetorical devices which closely recall the language of the Pauline account. The language of Protestantism is intimately related to choices made by Tyndale and Coverdale which concern the directness of the Bible.[7]

Bale's extensive biographical account of British writers in his *Summarium* (Wesel, 1547), expanded in his *Catalogus* (Basel, 1557 and 1559), provides another clue. This has ancient precedents in a type of work begun by St Jerome, and significantly extended by Gennadius in the fifth century.[8] Thus his writing before his conversion, especially in relation to the history of the Carmelite order, established a pattern which when he came to work for Leland in the later 1530s he could adapt in his manuscript work the *Anglorum Heliades*.[9] Here appears an early version of his autobiography, in the form of a full life in one page. We can see that it offers a strong sense of his own special fortune in being selected by God and in receiving instruction from Lord Wentworth in his conversion. This contrasts with the autobiography in *Vocacyon* because it actually refers to conversion, which became a key trope in Protestant and more especially Puritan autobiography.[10] The account in *Anglorum Heliades* was expanded in the *Summarium* and the *Catalogus*. In the latter he speaks of the contrast between his old life and the new:

> at Cambridge I wandered in complete barbarism of scholarship and blindness of mind...And immediately I was carried by the divine goodness from the arid mountain to the flowery and fruitful valley of the gospel: where I discovered that all things were built not in sand but upon solid rock.[11]

Bale's lyrical tone gives no hint of any of the agonised self-examination which became characteristic of the many Protestant conversion autobiographies.

By the time he came to write *Vocacyon*, Bale was ready to ignore the conversion and to concentrate upon the ministry, as indicated in his title and his echoes of the Pauline journey protected by God. Travelling offered the most frightful hardship to Paul as it did to Bale. Bale's summary is modelled on Paul's at 2 Corinthians 11:25–7. Here is Tyndale's version:

> I was once stoned. I suffered thrice shipwreck. Night and day I have been in the deep of the sea. In journeying often: in perils of waters: in perils of robbers: in jeopardies of mine own nation: in jeopardies among the heathen. I have been in perils of cities, in perils in wilderness, in perils of the sea, in perils among false bretheren, in labour and travail, in watching often, in hunger, in thirst, in fastings often, in cold and in nakedness.

This is matched by the cumulative rhythms of Bale's more extensive account:

> I have bene in journayes and labours, in injuryes and losses, in peines and in penuries. I have bene in strifes and contencions, in rebukynges and slaunderynges, and in great danger of poyseninges

and killinges. I have bene in parell of the heathen, in parell of wicked prestes, in parell of false justyces, in parell of trayterouse tenauntes, in parell of cursed tyrauntes, in parell of cruell kernes and galloglasses. I have bene in parell of the sea, in parell of the shypwrack, in parell of throwynge over the boorde, in parell of false bretherne, in parell of curiouse searchers, in parell of pirates, robbers and murtherers, and a great sort more. (124–36)

It is striking that the echo of the style is as powerful as the parallel in the items of the lists.

Nevertheless an apparent factuality is shared by the short but full-life accounts and the autobiographical narrative section in *Vocacyon*. In the years before Bale wrote this account, he had made the acquaintance of John Foxe, the martyrologist. This seems to have occurred around 1548 at the house of the Countess of Richmond in London. At this point Foxe must have been working on the prototype of *Acts and Monuments*, viz. the *Commentarii Rerum in Ecclesia Gestarum* which appeared in 1554. Foxe's work is notable for its inclusion of documentary evidence such as depositions when under trial. This may be something Foxe adapted from Bale, whose work on Oldcastle and Anne Askew had been presented in a documentary manner. The papers of the latter, Bale claims, are in her own hand, and it is remarkable that Foxe used them as well in his account. All this leads to Bale's support for his vocation, for he includes the letter of appointment from the Clerk to the Privy Council (50).

The structure of *Vocacyon* is one of its salient rhetorical features. Of its 112 octavo pages, 52 are concerned with the narrative of the journey to Ireland and the subsequent flight: the rest make up the Preface and the Conclusion. In these sections the narrative is by no means absent, but it is referred to in what one might call a thematic way, determined by rhetorical purposes. In the Preface Bale explains that he has three object-ives: to demonstrate the true role of a bishop, to show the persecutions to which the honest Christian is subject, and to reveal God's deliverance. In the Conclusion he adds a special warning to the people of England who may have been responsible for the suffering which has now come upon them in the form of Mary and her bishops (Gardiner again). He makes further mention of the danger of celibacy and the extent of the conspiracy he had encountered in Ireland. But his final note is the reassertion of deliverance as God purifies his people through suffering.

These themes are given very close attention in the central narrative as he shows the brazen abuses of celibacy as practised by the clergy in Ireland and in Cornwall, where he found things just as bad. It has been suggested that Bale's model here was the satirical edge found in the medieval sermons.[12] I suppose there is no doubting his word and that he did visit Cornwall, driven there, he claims, after a storm of Pauline proportions, but the interesting thing from the point of view of the argument is that he could hardly have a chosen a region where the old religion hung on so firmly: this has a rhetorical significance and indeed convenience far beyond its factual importance.

Interwoven with these possible models is the important aspect of the value Bale put on the role of preaching. This has three notable effects. He sees preaching as the primary task of a bishop in contrast to the worldly preoccupations of his enemies. Second, he constructs the narrative so that it carries forward in an exemplary fashion the doctrinal aspects which are at the heart of his teaching. Third, he weaves into the narrative what are practically verbatim accounts of some of the sermons he preached at key moments. Examples are in his attempt to direct the people in the way of true repentance as soon as he took up his position in Kilkenny cathedral (784–800), and in his sermon on St Bartholomew's Day just after Queen Mary was proclaimed when he took a Pauline theme: 'I am not ashamed of the Gospel' (Romans 1:16) (983–1011). This was a day of rejoicing for the local clergy which Bale sought to off-set both by his sermon and by the revival of three of his biblical plays. Thus there is in this account through his interest in the sermon both a content and a method.

As to the narrative itself, though the account is concentrated into the central part of the work, it has two remarkable characteristics. He does not hesitate to bring in polemical details, some from his preaching mode when it suits him. Overshadowing this, however, is the directness of the account of events which he portrays. Whether by accident or design, some of what he writes has a strong sense of actuality, as at the end of a piratical chase of two vessels

> they toke one by reason that his topsaile brake, and that was a shippe of Lynne. In this had they nothinge but apples, for he went for his loadinge. (1543–6)

And yet so often we find that details are there not so much because of what may be called a *res gestae* type of autobiographical method, but because their significance can be made to support the effects he is striving to achieve concerning his beliefs and his message.

To illustrate this method, which I believe is fundamental to the process of the autobiography, I should like to mention two episodes.

First, at the beginning of this momentous year, 15 August 1552, Bale recounts how on the first day of his recovery from a 'mortall ague' he learned that the King had come to Southampton so he took his horse 'about 10. of the clock, for very weaknesse scant able to sytt hym'. By two o'clock he had covered the distance of 5 miles and stood in the open street. John Philpot, a friend of his at Court, noticed him and the attention of the King was drawn to Bale. The King

> marveled therof for so much as it had bene tolde hym a lytle afore that I was bothe dead and buried. With that hys grace came to the wyndowe and earnestly behelde me a poore weake creature as though he had had upon me so symple a subject an earnest regarde or rather a fatherly care. In the same very instaunt as I have bene sens that tyme credibly informed his Grace called unto him the lordes of his most honourable counsell so manie as were than present willinge them to appoint me to the bishoprick of Ossorie in Irelande. (612–22)

One might question that phrase about being credibly informed; choosing bishops from forgotten clergymen who walk the streets seems a highly unlikely royal strategy, even if the young Edward was at the time increasingly able to make his own decisions. The point of the narrative, apparently so circumstantial, is really to establish Bale's loyalty to the King – important for his upholding of royal authority in Ireland later in the story – and to be as specific as possible about the King's direct intervention. There may also be an urge here to underline his own innocence and vulnerability, notably in 'fatherly care'.

The second brief episode concerns the murder of five of Bale's household servants on the 8 September 1553 who were haymaking on a Sunday 'betwixt viii and ix of the clocke after they had served God accordinge to the daye.' He tells how 'more than a score leaped out of their lurkynge busshes with sweardes and dartes and cowardly slewe them all unarmed and unweaponed without mercye.' Bale goes on to explain that 'they had watched so long afore, yea an whole month space they say'. The episode exemplifies the ruthlessness of his enemies and the growing conspiracy against Bale. It also serves to show that Bale and his people knew the right way to behave on Sunday, even if others did not. He explains that the incident was so frightful that the Suffren of Kilkenny sent no less than a 100 horsemen and 300 footmen to escort Bale to the safety of Kilkenny the same day (1142–65).

One of the interesting rhetorical aspects of the autobiographical narrative is the way in which Bale acts as an interlocutor. There are a number of conversations in which his own attitude is either stated or implied, and in looking at these one is often reminded of dramatic dialogue. At a critical point in the journey when the rapacious captain was threatening to sell Bale at Dover to the English authorities, Bale upbraided him

'I pray you therefore in your conscience that ye tell me what evil ye know else by me that ye make so terrible doings.'

'I can not see,' saith the Captain, 'that ye will be ordered after any good sort.'

My only misorder was then that my money was in my purse and not in his. Whereunto I answered with an heart full of dolour and heaviness to behold men's so damnable practices of mischief for filthy lucre's sake: 'I am contented, master Captain,' said I, 'to be ordered as ye will reasonably have me.'

'What will ye give, then,' said the Captain, 'to be delivered into Flanders and our purser called again?'

I answered that I would give as his reason and conscience require.

'If ye had told us so much yester night,' said he, 'this matter had been at a point, and we by this time had been in Zeeland.' [13]

In another significant passage of dialogue after dinner at St Ives, Bale and a gentleman complicit with him are quizzing Sir James, a priest, about sexual and other irregularities. The gentleman, reports Bale, 'rubbed me on the elbowe and bad me in mine eare to lete him alone and I shulde heare wonders' (1472–4).

Occasionally, he shows himself out to score off his opponents as with a boatman whom he outwitted. Bale comments then 'went he awaye in great displeasure with no small reproches' (1521). A similar motive appears in the examination before the Bailiff at St Ives. When the Bailiff asked his accuser, Walter, how long he had known Bale and what treason he knew Bale to have committed,

He answered that he never saw me neither yet had heard of me afore I came into that ship of war a 3 or 4 days afore.

Then said the bailiff 'What treason hast thou known by this honest gentleman since? For I promise thee he seemeth to be an honest man.'

'Marry,' said he, 'he would have fled into Scotland.'

'Why,' said the bailiff 'and knowest thou any impediment wherefore he ought not to have gone into Scotland?'

'No,' said the fellow, 'but he was going toward Scotland.'

'If it be a treason,' saith the bailiff, 'to go towards Scotland, a man having business to do there, it is more than I knew afore. And truly,' saith he, 'then are there many traitors abroad in the world! Good fellow,' said he, 'take heed that thy ground be good in accusing this man, else art thou worthy to suffer due punishment for it. For thou doest it else upon some other affection than desire of right.'

With that he stood still and was able to say nothing for he was as drunk as an ape. (1427–43)

Of course, Bale never claims outright that he was virtuous: indeed on the contrary he describes himself at one point as 'but a clodde of coruption, feeling of myself as of my self nothinge els but sinne and wickednesse' (247–8). One might of course reasonably ask whether these passages are factual accounts of what actually happened or imaginative reconstructions. The 'fiction' is intriguing.

Part of the rhetoric of the *Vocacyon* lies in some of the technical details surrounding the presentation of the book itself. According to the colophon it was 'Imprinted in Rome before the castell of S. Angell at the signe of S. Peter in Decembre Anno D.1553'. Rome is a pointed impertinence, but the dating is also suspect because Hugh Singleton, the printer whose mark appears on the volume, was in prison in London until after the end of 1553.[14] From the dates Bale gives, perhaps from a journal, it is clear that he left Dublin around the middle of September. The pirating took more than a week before Dover, and then another 23 days more before they reached the Netherlands. He cannot have started the book much before the end of October, which does not give him much time if it were to come out by December. Probably he wanted to make it seem as though it were a quick response, and to bring as much pressure as he could to bear on the developing situation in Marian England.

The book itself has illustrations which show the conflict between the English Christian and the Irish Papist (frontispiece) and a picture of an allegorical truth carrying the 'Verbum Dei'. Bold type is used in lists in the texts for the names of virtuous heroes of various sorts, and for scriptural quotations. There is also a fine craft in the annotations which appear in the margins. These often set off the polemic. One passage about sexual excesses in the clergy has successively 'Occupienge...priapus...Celibatus' (1871–84). Even the index carries meaning, as in 'Thomas Hothe, a wicked justice' and 'vocacion of the Authour, just'.

Most of these extratextual devices may be found in Bale's other polemical works which helps to place *Vocacyon* in a wider context: but the function here is closely related to the presentation of the self as part of the management of the book.

*

Having discussed, all too briefly, some of the rhetorical aspects of this work, I want to suggest that it evidences a high degree of literary sophistication on Bale's part. This is supported particularly, I think, by the firm structure with the autobiographical narrative forming the dominating central passage and the clever exploitation of it in advance and in retrospect. Perhaps the most interesting aspect is the way in which Bale organises his experiences and presents his self. Notably there is little self-interrogation, and very little sign that Bale needs anyone to boost his self-confidence. We find Bale not so much by what he says directly about himself as from his actions which are often described without any comment to indicate his own feelings, though the implication that he was acting in an ideologically sound way is never far from the surface. Where he does attempt a definition of self, Bale achieves it by historical and theological issues, by his need to persuade, by the nature of his vocation and his role as a bishop, and by his persecution.

And yet, though it may be incomplete, a kind of personality does emerge, albeit one to be observed from outside. Pinpointed in the conversations we see him as obedient and righteous. At times his silence speaks volumes as he allows other people to reveal their weaknesses and to condemn themselves. In this, I suggest, his authorial stance may remind us of his earlier skill as a dramatist who can make his feelings and attitudes clear to his audiences by the words of others, even though he himself remains silent. There is also an ingenuousness which turns aside the presumption of comparing himself to St Paul, and indeed, in undergoing suffering, to Christ himself. Bale's enemies are called Pilate and Judas, reflecting an intention to be as close to Christ as possible.

But perhaps in contrast to all this, and somewhat inconsistent with the theological purposes which he earnestly pursues, there are hints of other interests. He wants to show that he has gained no worldly benefit from his adventures, and he makes several references to his own poverty. Another aspect is a concern for detail in its own right, even though this may ultimately be linked with the need to witness the divine truth. Besides particular episodes like the sea chase, this shows itself in the documentary evidence, and the scrupulous use of dates – there are about twelve of them – suggesting either a phenomenal memory or a lost journal. There are plenty of other, external reasons for supposing that Bale rarely had a pen and notebook out of his hand.[15] There is also the possibility that he was able to refer to notes or texts for the sermons he preached during these momentous events.

With regard to the aspects of autobiographical writing highlighted in the introduction to this present volume, it is perhaps clear from my discussion that the form of the *Vocacyon* is largely generated from Bale's extensive production of polemical pamphlets. These used a rich texture of biblical example, and took an outstandingly forceful line against his opponents. But the clarity and immediacy of many of the details of the central narrative section show a preoccupation with actuality even though the inclusion of such details is part of the polemic. In addition, Bale's sense of the self seeks ancient and historical authority from the Pauline journeys and the biographies of early Christians.

Bale's conception of his role and function as a bishop is an integral part of his sense of himself as an example. As he means to challenge what he considered to be the wrongs of his papist counterparts, his purpose is controversial and destructive, but there is a positive side to this because it shows his readers how a bishop should conduct himself in adversity. Moreover the personal side is often, as we have seen, presented with considerable realistic detail, but without direct comment, especially in the quasi-dramatic conversations incorporated to show how he had prevailed. If we add to this Bale's intense presentation of himself as persecuted for Christ's sake – a feature shared with his other writings[16] – we find a paradoxical combination of submissiveness and triumph. One has an impression that he wants to display the purity of his religious stance, and that he knows that humility is more desirable than boastfulness.

The exemplary aspect is intimately linked with Bale's motivation. The exchanges where he is proved right, and his success in outwitting his opponents or revealing their shortcomings, may give a rather unattractive impression of superiority. He may seem argumentative and self-righteous at times. Yet he has to present this as part of his mission since it reflects the power of the divine influence which he believed supported him. His created self is not simply a matter of self-indulgent vanity, but a function of his inspirational mission. He does not ask for pity, but rather suffers that God's care may be manifest. This is powerfully sustained by his belief that, like St Paul, he had been protected by a special grace.

Finally, after experiencing a year like this one of Bale's I think most of us would want to write something down: whether to boast, to be thankful for having survived, to glory in the role of sensational story-teller, or simply to come to terms with a whirlwind of events which, again and again while they were unfolding, threatened to be over-whelming. Possibly this psychological necessity enables us to decon-struct the intense commitment to a religious message and I think it makes for an insight into what we might perceive as a necessity for 'autobiography'.

Notes

1   References are to *The Vocacyon of Johan Bale*, ed. Peter Happé and John N. King, Renaissance English Text Society of America (Binghamton, 1990).
2   Jerome Bruner, 'The Autobiographical Process', in *The Culture of Autobio-graphy*, ed. Robert Folkenflik (Stanford, 1993) p. 39.
3   Barrett J. Mandell, 'Full of Life Now', in *Autobiography: Essays Theoretical and Critical*, ed. James Olney (Princeton, 1980) pp. 49–72.
4   See *The First Examinacyon of Ann Askewe* (Wesel, 1546), sig.C7^r. *The Lattre Examinacyon of Anne Askewe* followed in 1547.
5   *The Examinacyon and death of the blessed martyr of Christ Syr Johan Oldecastell* (Antwerp, 1544). For Askew see note 4 above.
6   Paul Delany notes that by the seventeenth century 'autobiographers com-monly distorted the true pattern of their lives by trying to fit every detail into the Pauline archetype': *British Autobiography in the Seventeenth Century* (London, 1969) p. 30.
7   For a discussion of this language, see John N. King, *English Reformation Liter-ature* (Princeton, 1982) pp. 138–44. But the value of rhetoric was noted by Nicholas Udall: 'For divinitie, lyke as it loveth no cloking, but loveth to be simple and playn so doth it not refuse eloquence, if the same come without injurie or violacion of the truth' (quoted on p. 141).
8   See my *John Bale* (New York, 1996) pp. 63–4.
9   BL Harley MS 3838, c.1536.
10 Delany, op. cit., pp. 33–4.
11 *In omni literarum barbarie ac mentis caecitate illic et Cantabrigiae pervagabar . . . Protinusque divina bonitate, ab arido monte in floridam ac foecundam Evangelii vallem transferebar: ubi omnia reperi non in arena, sed supra solidam petram aedificata. Catalogus,* I, 702; my translation.
12 Leslie P. Fairfield, *John Bale: Mythmaker for the English Reformation* (West Lafayette, 1976) p. 335.
13 For this and the following example I have modernised spelling and layout in order to set off the immediacy of the conversations.
14 Christina H. Garrett, '*The Resurrection of the Masse* by Hugh Hilarie – or John Bale', *The Library*, 4th series, 21 (1941) 143–59.
15 One of his most remarkable feats was that he sustained and apparently carried about with him through many journeys and vicissitudes a massive amount of bibliographical information: this must have started in the 1520s and continued until the compilation of the *Catalogus* in the 1550s. Several of his extant manuscripts reveal the vigour and intensity of this work.

16 'I have geven myselfe over unto poverte, and unto peynefull exyle with my
wyfe and chyldren, and schall not, I trust, refuse the deathe also yf yt come
that waye. For so necessarye ys yt now to suffre for Christes doctryne as in
the aposteles tyme,' *A Dysclosynge or Openynge of the Manne of Synne* (Antwerp, 1543) p. 6. 'The trewe churche of Christ knowne...but by persecu-
cyon for ryghtousnesse sake,' *Yet a Course at the Romish Foxe* (Zurich, 1543),
fol. 34v.

4
Songs, Sonnets and Autobiography: Self-representation in Sixteenth-century Verse Miscellanies

Elizabeth Heale

The term 'autobiography' must come hedged with numerous qualifications when we are dealing with the sixteenth century. Terms such as 'identity' (in the sense of a distinctive personality) and 'individuality' do not begin to be used until the seventeenth century, while 'autobiography' and 'individualism' are nineteenth-century coinages.[1] As the introduction to this volume points out, we can no longer accept Burckhardt's famous pronouncement that man emerged in the Renaissance as a fully self-conscious 'spiritual individual' after a period in which he 'was conscious of himself only as a member of a race, people, party, family, or corporation'.[2] Such a simple model has been challenged on the one hand by those medievalists and historians who point to clear evidence of individualism and self-consciousness long before the Renaissance, and on the other by theorists who challenge all notions of a unitary essential self waiting to spring into full consciousness of its being.[3]

My own contribution to the discussion of early modern autobiographical writing examines a new interest in lyric self-dramatisation and the use of personal autobiography in mid-sixteenth-century miscellanies authored by such little discussed writers as Barnabe Googe, Thomas Howell, Isabella Whitney, George Turberville and the better known George Gascoigne. Although little read in the twentieth century, the printed poetry of these writers had a formative influence on later Elizabethan poets, perhaps particularly encouraging the development of autobiographical poems and sequences, a taste fed from another direction by the continental fashion for Petrarchan sonnets.[4] I argue that there are two crucial factors in producing the distinctive self

59

dramatised through the mid-century miscellanies: one is the availability of print and its transformative effect on early Tudor courtly and coterie poetry; the other is the particular circumstances of the poets who, through birth or education, claimed gentility, but who nevertheless had to sell themselves in a competitive marketplace.

A tendency to conflate persona and author may always have attended first-person love lyrics, and has often been deliberately exploited, not least by Petrarch in the *Rime Sparse*.[5] I would argue, however, that print subtly altered the potential significance of personal allusions. In an environment of elite manuscript circulation and the shared reading and social enjoyment of courtly lyrics, autobiographical allusions may be understood as primarily performative, contributing to the pleasure of a familiar circle, or designed to make a political or moral point. Many first person lyrics that in print seem to be intimate and self-expressive documents were, in a manuscript system, shared and copied, sung, adapted and imitated, the name of the originating author often becoming a matter of legend or indifference.[6] Many of Petrarch's passionate sonnets, for example, were set to music by trained musicians and performed for the pleasure of elite audiences, while some of the apparently confessional and personal poems of Wyatt and his contemporaries were adapted for singing by three or four voices to the tunes of popular dances.[7]

The packaging and distribution of lyrics in the mid-sixteenth century, in printed miscellanies, to be bought and read, largely silently, by individuals in the privacy and isolation of hired lodgings and bedchambers, rather than shared in the withdrawing rooms of palaces, provided an environment in which the first-person voice of the lyric could be heard in a new way, as primarily private and confessional, whispering the passionate experiences of the author to the inner ear of the reader. The printing of courtly lyric poems produced new kinds of readers, and helped to shape new kinds of poems and poetic voices. These first become apparent in the series of mid-Tudor single-author miscellanies. The manuscript 'book of songs and sonnets' by Thomas Whythorne, written in manuscript in *c*.1575–7 at the height of the miscellany fashion, to which I shall turn at the end of this essay, suggests that the 'autobiographical' lyric voices of the printed miscellanies could in their turn help to shape a contemporary's intimate narrative of own life.[8]

The mid-century authored miscellanies are crucially shaped by the first major printing of manuscript English lyric poems, the *Songes and Sonettes*, issued by the printer Richard Tottel in 1557. So immediately popular was this anthology, commonly known as *Tottel's Miscellany*, that it went through three separate printings in the first year of its appearance, with at least nine editions appearing before the end of the century.[9] Without *Tottel's Miscellany* it is difficult to imagine the mid-century miscellanies existing or taking the form they did.

Tottel's project was to make public the private poems of courtly and gentlemanly amateurs previously circulating in manuscripts owned by, in Tottel's ironic phrase, 'the ungentle horders up of such treasure' (I. 2). As we hold Tottel's small quarto in our hands, our sense is not of private matters drawn forth onto the public stage, but of privileged access into a cultivated private world. The volume gives us the illusion that it has merely extended the process of manuscript copying to reach a wider network of privileged readers, some of whom seem to have treated it as they would a manuscript miscellany, answering, adapting and freely imitating individual items.[10] Those who bought *Tottel's Miscellany* were given access to the social and political verse of 'the noble Earle of Surrey and... the depewitted sir Thomas Wyat the elder' and their like (I. 2). Part of Tottel's purpose in printing such poetry, he tells his readers, is to encourage us to imitate our betters, both their country verse and their courtly refinement. thus 'the unlearned [might] ... learne to be more skilfull, and to purge that swinelike grosseness, that maketh the swete maierdome not to smell to their delight' (I. 2).

As Wendy Wall has suggested, *Tottel's Miscellany* and its imitators 'marketed exclusivity ... [they] functioned as conduct books ... because they demonstrated to more common audiences the poetic practices entertained by graceful courtly readers and writers.'[11] Such imitation was not simply didactic, or a matter of snobbery; it could, more crucially, be a matter of finding a job. Daniel Javitch has observed that 'until the 1580s Elizabethan writers seeking administrative jobs at the Queen's court used their verse and prose to exhibit their humanist learning and oratorical proficiency in the hope that such talents would be put to pragmatic use'.[12] Whether at the Queen's court or in the households of noble, or upwardly mobile, families, an ability to behave and write like a gentleman, with courtly graces and a well-stored head, was an important asset. Through imitation of *Tottel's Miscellany*, mid-Tudor poets could fashion themselves to a wider public of prospective employers and patrons as personable, accomplished and flexible employees. As Whythorne explained when printing his own *Songes* in 1571, publishing could make the author 'to be known of many in the shortest time that might be'.[13]

The poems in Tottel are by many hands and are the products of diverse occasions over many years, but their collection within one volume had the effect of 'marketing' a composite courtly behaviour, comprising both the graceful passion of the lover and the 'deep-wittedness' of the sage counsellor. Early readers of Tottel seem not to have been particularly attentive to the different identities of contributors. The book's title-page announced that it contained *Songes and Sonettes, written by the ryght honorable Lorde Henry Haward late Earle of Surrey, and other* (*sic*) and it seems that for many readers, all the poems were, in effect, Surrey's, and read as the varied performances of the ideal nobleman/ courtier represented by Surrey.[14]

Tottel's lyrics typically develop topics of intimacy, retirement and privacy. A number of Surrey's own poems dramatise a first-person voice of private integrity and introspection set in opposition to, or alienated from, the world at large.[15] Throughout the *Miscellany* many poems return to the themes of the dangers and corruptions of courtly service and the virtues of retirement, or the 'meane' (i.e. middling) estate.[16] A central rhetorical trope of poems of good counsel, such as those on the mean estate, is the mutual authorisation of ancient wisdom and personal experience.[17] Thus Wyatt and Surrey sometimes explicitly draw on aspects of their own lives to reinforce truisms in their poems. Wyatt's epistle, 'Mine Owne John Poyntz', for example, personalises its conventional message of the corruption of courtly ambition and the pleasures of retirement (translated from the Italian of Luigi Alamanni) by adding strategic allusions to his own circumstances in 1536.[18]

The ironies and political manoeuvring inherent in a poem about virtuous retirement by a courtier as ambitious as Wyatt are easily lost when the poem is removed by print from the milieu for which it was written. In Tottel, Wyatt's artful epistles become elegantly transparent vehicles for the themes signalled in their titles: 'Of the Courtiers life' or 'Of the meane and sure estate'. Tottel's titles announce the topic or, particularly in the case of the amatory verse, the supposed occasion: 'The lover sheweth how he is forsaken of such as he sometime enjoyed', 'The lover hopeth of better chance'. Paradoxically the effect of such titles is to erase all sense of specificity or social wit; their sentiments are explained simply as the transparent expressions of sage counsel or a conventionalised lover.

In emphasising the sentimental or moral contents of the poems, Tottel's titles tend to erase their artfulness and focus our attention on the expressiveness or feeling of the speaker. This latent 'autobiographical' effect is further encouraged by an easy slippage between references to individual poets as 'he' or 'his', and the conventional 'lover' of many of the titles. Thus a sonnet by Surrey entitled 'Description and praise of his loue Geraldine' is preceded by one entitled 'Complaint of the louer disdained' and followed a few poems further on by 'Prisoned in windsor, he recounteth his pleasure there passed', an explicitly autobiographical poem (I. nos. 7, 8 and 15). It is uncertain whether the 'he' of the Geraldine sonnet should refer to Surrey or to the conventional 'lover' who figured in the previous poem. The distinction between the historical Surrey and a conventional lover may not be important for Tottel who is more concerned with the exemplary function of his courtly poems, but evidence that a slippage did in fact take place, encouraging in readers an 'autobiographical assumption', is well illustrated by Surrey's Geraldine sonnet.

The addressee of the Geraldine sonnet, 'From Tuscan cam my ladies worthi race', can be identified as Lady Elizabeth Fitzgerald, a girl who may have been only nine years old when the poem was written.[19] Nothing more is known of any relationship between Surrey and this young girl and the poem is almost certainly a playful courtly compliment. An embryonic tendency in the *Miscellany* to elaborate this single sonnet into a romantic narrative may explain the appearance, in the second edition of Tottel only, of a form of her name 'Garrat' in a second sonnet, 'The golden gift.[20] Much later in the century, such hints were spectacularly developed by Thomas Nashe in his novel *The Unfortunate Traveller, or The Life of Jack Wilton* (1594) which depicts a ludicrously fictive Earl of Surrey travelling to Florence to defend the honour of Geraldine's beauty against all comers. The love-sick Surrey is described penning spontaneous 'extemporal ditties' (of Nashe's composition) to Geraldine.[21] He appears at the tournament in her honour dressed in a fantastic armour which bespeaks his love-sick condition:

> the bases thereof bordered with nettles and weeds, signifying stings, crosses and overgrowing encumbrances in his love; his helmet round-proportioned like a gardener's water-pot, from which seemed to issue forth small threads of water, like cittern strings, that not only did moisten the lilies and roses, but did fructify as well the nettles and weeds, and made them overthrow their liege lords.[22]

The story was taken up more soberly by Michael Drayton in *Englands Heroicall Epistles* (1597–9), two of which relate to an affair between Surrey and Geraldine. Surrey is imagined writing to Geraldine from Florence, once again composing verse from an overflowing heart and carving his lines (of Drayton's composition this time) on the trunk of a tree.[23] In his notes to the poems, Drayton claims that a number of Surrey's poems from *Tottel's Miscellany* refer to his love for Geraldine. Even Wyatt's 'Tagus, farewell' is tentatively purloined as evidence of the sentimental patriotism of his semi-fictional Earl.[24]

In offering exclusivity to a new audience of aspiring gentlemen and courtiers, Tottel sold a very specific notion of what it was like to be cultivated and 'gentle'. The accomplished courtier must both be, and express himself elegantly as a lover, a loyal friend and a wise man. In the series of mid-century authored miscellanies, a middling class of aspiring clients, gentle by birth and adequately educated, but in need of appropriate employment, attempted to make themselves 'to be known of many in the shortest time that might be' through the medium of print and in imitation of Tottel's prestigious example.

It is impossible in the space available to examine all the miscellanies in detail, so I shall focus on one example and draw analogies with others. The collection 'by Thomas Howell, Gentleman' called *Newe Sonets and Pretie Pamphlets* is in many ways typical. It survives in an undated revised edition 'newly augmented, corrected and imprinted', licensed to Thomas Colwell in 1567–8.[25] Another miscellany by Howell, *The Arbor of Amitie. Pleasant Poems and pretie Poesies*, was printed shortly afterwards in 1568.

Many individual items in *Newe Sonets* derive from Tottel, and he also carefully follows the *Miscellany*'s mixing of love poems and poems of moral reflection. Howell is keen that we should be aware of his gentlemanly status and he advertises it not only on the title-page, but in the various clear signals about his background and his participation in the manuscript exchange of poems by cultivated men. In a dedication to 'Master Henry Lassels Gentilman' he describes his poems as 'trifling toyes' designed simple to amuse his friends and printed now only at their 'earnest request'.[26] Howell's own name figures prominently in the volume, particularly in a number of poems exchanged with his friend John Keeper. For example, at the end of *Newe Sonnets* features a poem on 'The Unsertaintie of seruice by John Keeper to his friend Howell', which warns of the perils of service at court. This is answered by Howell 'to his friende Keper' on the virtues of service to God. In turn, Howell's poem is answered by Keeper (pp. 153–9). Howell is here publicising his part in a discriminating and cultivated friendship. In another exchange, printed in the 1568 *Arbor*, John Keeper bids farewell to Howell, who is reluctantly departing 'To set thy selfe in sounder sort'. Howell's name is worked into a line,'And stil I *Ho* to see him *well*' (pp. 94–5). Like Tottel, Howell provides titles indicating the subjects and supposed occasions of his poems: 'The description of his loathsom life, to his friende', 'The Louer almost in desperation, moueth his estate' (pp. 122–3). As in Tottel, the identity of an apparently conventional 'lover' threatens to elide, through the uncertain reference of 'he' and 'I', into the voice of the poet, in this case the Howell who figures so conspicuously in his own volume.

The temptation to read autobiographically is reinforced by the first poem in *Newe Sonets*, entitled 'He declareth his greate mishappes, and lamentable sorowes of harte' (pp. 117–19). This poem owes much in form and diction to a number of medium-length poulter's measure complaints in Tottel.[27] Howell's complaint, however, is not against love but against ill-fortune which has deprived the speaker of the means to sustain the privileges of his upbringing:

> Would God when I began, to enter first in life,
> That present death had pearst my hart, and rid me cleane this strife.
> So should my Parents not, haue been at such great cost,
> To bringe me vp on whom by fate, their great good gifts are lost:
> Ne yet haue left to me, no whit such wealth at all,
> Whereby from wealth to miserie, might chaunce a soden fall.
> (pp. 117–18)

In some respects this is very conventional – the turns of fortune's wheel figured repeatedly in medieval literature as in *Tottel's Miscellany* – but the specificity of the poem's narrative of economic disappointment, and the speaker's complaint that the mean estate is hard to bear by someone used to something better, signals a new kind of voice, identified as Howell's own, asserting a social perspective which is unlike Tottel's contributors. The poem serves, at the head of *New Sonets*, to place Howell socially with some precision. Above all, Howell laments the loss of personal liberty, a detail that lends sharpness to the advice in John Keeper's later poem, addressed to Howell, on 'The Unsertaintie of seruice' in which Keeper warns:

> Now master calls, now mistris speakes, now vp and downe goe now,
> now tarie here, now goe thou theare, at all commaundes be thou.
> Yet when thy maister likes thee well, thy mistres may thee hate:
> and thus betwene Caribdis rockes, thou sailst in doubtfull state.
> (pp. 154–5)

Thomas Howell became a 'ladies's man-servant'.[28] Howell's second volume, *The Arbor* (1568) addresses Lady Anne Talbot in whose 'daylie presence' Howell now is (p. 6). A later volume, *H. His Deuises* (1581) identifies the poet as the Countess of Pembroke's 'humble and faythfull Servant' and indicates that the poems were written in her household (pp. 165–6). If the first volume, *Newe Sonets*, was a gambit of self-promotion, advertising Howell as the gentleman he claimed to be, in manners and accomplishments as well as in birth, then it seems to have been successful, perhaps playing a part in his eventual employment by the literary Countess of Pembroke. Nevertheless, even as the volume served Howell's career, as it was designed to do, it also betrays, in an entirely new kind of autobiographical poem, the pains and frustrations of his middling condition, bred to perform with cultivated ease the gestures of a class for which his income was inadequate. For Howell the only prospect was to lose his gentlemanly liberty and become a despised, but courtly, serving man.[29]

Howell's autobiographical poem introduces what becomes a distinctive voice in the mid-century authored miscellanies, that of a cultivated, but economically dependent, speaker who, while demonstrating his mastery of the manners and gestures of a social elite, records his personal misfortune and experience of social marginality. Of course the motive in publishing such personal histories is self-promotional. They are designed to appeal to the sympathies and tastes of generous patrons. Nevertheless, out of the prestigious materials of *Tottel's Miscellany* the mid-century poets constructed a new kind of autobiographical voice, telling a new and significantly different narrative of social aspiration and threatening failure. Both in their volumes as a whole, and in individual poems of personal history, Tottel's imitators created a new form for self-fashioning, and plaintive self-assertion.

The forms such autobiographical self-assertion took were highly innovative. Barnabe Googe and George Turberville developed the first embryo amorous sequences in English by mingling details of their careers with love poems.[30] Turberville, for example, interweaves references to an expedition he made to Russia in 1569, in the entourage of the English ambassador, with love lyrics telling a narrative of separation and uncertainty.[31] Gascoigne's *A Hundreth Sundrie Flowres* (1573) plays with particular skill on his paradoxical need both for self-promotion and gentlemanly privacy. His miscellany collection pretends to be a manuscript album by many hands printed without permission by one G.T. Gascoigne's name, however, appears with striking frequency. The fiction of a pirated manuscript allows Gascoigne to present his poems as particularly prized items in a private gentlemanly collection, as Surrey's had been among the anonymous 'uncertain authors' of Tottel. Among the poems explicitly ascribed to Gascoigne is 'Gascoigne's Woodmanship', an apparently autobiographical account of the poet's career, addressed to Lord Grey, and set in the context of a shared day's hunting on Lord Grey's estate. It tells the story of Gascoigne's life as a history of bad luck in spite of the poet's multiple efforts and extraordinary qualifications. The grace and humour of the tale's telling demonstrates the cultured return a generous patron would get on any investment. Even as Gascoigne advertises his need for substantial patronage, he elaborately displays himself as a courtly insider, participating in the elite practices of hunting and manuscript circulation.

Less well known than Gascoigne's tale of himself is Isabella Whitney's witty and graceful version of her life in her miscellany, *A Sweet Nosgay*, published in 1573, the same year as Gascoigne's collection.[32] Whitney demonstrates that the potential for self-display of the mid-century authored miscellany could be appropriated by an unusual woman to publish her own accomplishments, and possibly make a bid for patronage. Whitney's decision to publish her secular poems in the form of a miscellany was a remarkably bold act, but she mitigates the boldness by altering her self-presentation to conform in some respects to gendered expectations. Poems of courtly amorous dalliance, standard among imitators of Tottel, are omitted.[33] Instead, Whitney presents herself first, in a long series of virtuous verses derived, she reassures us, from the writing of a man, Hugh Plat.[34] However, her volume gradually becomes more self-assertive. In a middle section of personal verse epistles and poems, Whitney represents her self as central to a network of close, but far-flung family and friends to whom she writes with news and expressions of concern about their welfare. Whitney herself appears marginal to the social network she helps maintain. She is unmarried, without a household of her own, dependent, as she makes clear, on the decisions of family and employers for her place and manner of living. In the final poem of the volume she uses the conceit of a last 'Wyll and Testament' to London to express the ambivalence of her relationship with the city from which she is 'though loathe to leave . . . upon her friend's procurement . . . constrained to depart.[35]

The rather bitter joke of the poem is that not only can Whitney not choose where to live, but she has no wealth of her own to bequeath. Her bequests are 'such Goods and riches which she moste abundantly hath left behind her', that is, all the untouched wares of the city. In her poem she flits ghost-like through a London of shops and shopkeepers, as unaltered by her absence as they were unaffected by her presence:

> I goldsmiths leave with jewels such
> As are for ladies meet,
> And plate to furnish cupboards with

> Full brave there shall you find,
> With purl of silver and of gold
> To satisfy your mind;
> With hoods, bongraces, hats or caps
> Such store are in that street
> As, if on t'one side you should miss,
> The t'other serves you feat.
>   (ll.51–60)

Passing through Cheapside, Birchin Lane, St Martin's and Temple Bar, she comes at last to Ludgate prison, where she pauses:

> I did reserve that for myself,
> If I my health possessed
> And ever came in credit so
> A debtor for to be.
> When days of payment did approach,
> I thither meant to flee,
> To shroud myself amongst the rest
> That choose to die in debt.
> Rather than any creditor
> Should money from them get
>   (ll.179–88)

Whitney's vivid geographical evocation of the city and of the marginality of her female, penniless presence in it, contributes an unusual and eloquent example of the new kind of autobiographical poem that emerges from the mid-century miscellanies.

I want to conclude by looking at Thomas Whythorne's 'A book of songs and sonnets, with long discourses set with them, of the child's life, together with a young man's life, and entering into the old man's life'.[36] Although its title points to Tottel and the miscellanies as important models, this is not a miscellany, but an account of the writer's life as a context for his own poems. To some extent the narratives can be seen as explaining the occasions and the meanings of the poems in the same way as, but at much greater length than, the explanatory titles characteristic of Tottel and of all the mid-century miscellanies, particularly Gascoigne's. The manuscript of 'A book of songs and sonets' can be dated to *c*.1575–7, but it was not published until the twentieth century.[37] Whythorne, who was a professional musician, did however publish some of his own poems to accompany his musical settings in *Songes* (1571). Another collection of songs appeared in 1590.

The sheer length and detail with which Whythorne develops the narrative of his life makes his book unique, but the man that emerges from its pages in many respects is modelling himself according to the same ideals as the mid-century miscellanists. The manuscript's title, pointing on the one hand to the 'songs and sonets' of courtly dalliance and on the other to the moralised three ages of man, indicates Whythorne's sense of the need to perform, *à la* Tottel, as both a courtly lover and a sententious philosopher. The tensions of performance on both fronts, visible in most of the mid-century miscellanies in which the moralist shows considerable unease with the suspect passions of the lover, are particularly apparent in Whythorne's narrative. He gathers, in good schoolboy fashion, numerous sentences of aphoristic misogyny from his favourite 'author', the biblical book, 'Jesus Sirac' or Ecclesiasticus, in order to fortify his masculine virtue against the seductions of women (see, e.g. pp. 13–20 and 167–8). So formative in his early years are these sayings that he 'could... by no means content myself to brook and abide in the services of Cupid and Venus: it grieved me ever when I did hear thereof' (p. 13).

It is ironic that Whythorne should thereafter have spent most of his life, according to the 'book of songs and sonets', writing and performing amorous ditties for female patrons, all of which he represents as part of covert sentimental liaisons. Whythorne boasts of his success in the arts of courting which he defines as being 'in company with women, to talk with them, to toy with them, to jibe and to jest with them, to discourse with them, and to be merry with them (all the which some do call courting)' (p. 24). Various episodes, including one in which verses are left in his gittern strings, seem almost to mimic Gascoigne's erotic romance, 'The Adventures of Master F.J.', printed in *A Hundreth Sundry Flowers*, which similarly weaves a prose history around a sequence of amorous lyrics.[38] Whythorne's anxious efforts to behave both as a courtier and as a moral gentleman, produce, on occasion, comic results, as when he playfully leaves a verse in his mistress's (for Whythorne the term is always ambiguous) chamber. Whythorne's account carefully preserves his own compositions:

> *When pain is pleasure and joy is care,*
> *Then shall good will in me wax rare.*

This writing I left where I found the said implements to write withal; and coming the next day to her chamber, I found written as followeth:

> *For your good will look for no meed,*
> *Till that a proof you show by deed.*

Unto the which when I had seen it, I replied as followeth:

> *When opportunity of time serveth,*
> *Then shall you see my heart swerveth.*

After this I understood that the answer to my foresaid rhythm was
not made by my mistress, but by a waiting gentlewoman of hers, of
whom, although I needed not to have thought scorn, I repented me
of that which I had then done, because I like not to make love to two
at once. (p. 34)

The episode illustrates the combination of flirtatiousness and pro-
priety which characterises Whythorne's narrative throughout. It also
illustrates his uncertainty about his own social station. On the one
hand, as a professional musician in the household, he was, as he
admits, only 'an ace above' a serving man (p. 74), on the other, he was
employed to teach courtly skills. In the latter capacity, he was typically employed to assist mothers in bringing their children up 'gentle-
women-like' (p. 84). This serving man taught courtly skills. The
number of misogynist jokes and sayings throughout the manuscript
may have much to do with the perceived effeminacy and social stigma
of Whythorne's constant association with women and children. His
representation of his relations with female employers (his mistresses)
primarily in terms of a covert amorous courtship with them may serve
to reassure himself and his 'reader' of his own male potency.[39] Similarly,
his repeated reluctance to become the servant/lover he imagines his
mistresses want may stem as much from his sense of his own social
dignity and gentlemanly liberty as from his moral virtue. In response
to what he takes to be the amorous petulance of a rich widow, he
expostulates in his manuscript:

And I, for my part, seeing that my profession hath been and is to
teach one of the seven sciences liberal, the which is also one of the
mathematical sciences; and in the respect of the wonderful effects
that hath been wrought by the sweet harmony thereof, it passeth all
the other sciences; I do think that the teachers thereof (if they will)
may esteem so much of themselves as to be free and not bound,
much less to be made slave-like. (p. 46)

The terms that Whythorne uses to assert his threatened status are strikingly close to those of Thomas Howell when faced with becoming a lady's serving man.

For whatever reason Whythorne began his 'book of songs and sonets', it becomes in the course of writing an extremely intimate document, which Whythorne could not have published. The document reveals, I suggest, a private self moulded by the contradictory aspirations and assumptions fostered by Tottel and the mid-century miscellanies. For Whythorne lyrics are personal documents, transparently expressing events and emotions in his own life. On the other hand, Whythorne exploits the innuendo of courtly lyrics, their suggestion of private meanings and covert intrigues. For Whythorne they are signs of a courtly way of life that he both desires and fears, and to which he never quite feels he belongs.

Whythorne's uncertainty about the transparency of courtly lyrics is particularly evident in his agonising about the publication of his 1571 *Songes*. After a series of frustrations and disappointments in his career, he wanted to publish in order, he writes, to 'make myself to be known of many in the shortest time that might be' (p. 140). He sees the book and its contents as intimately related to himself:

And then came to my mind that, seeing the books with the music in them should be as my children, because they contained that which my head brought forth, (as it is said that the greatest substance, although not the chiefest, whereof children be procreate, cometh from the head and brains); also because they should bear my name, I could do no less than set in every one of them their father's picture or counterfeit, to represent unto those who should use the children the form and favour of their parent. (p. 175)

The book is his 'child' not only because he produces it, but because its contents are like him. He includes an engraving of himself to demonstrate the similarity. The self he displays through the printed book is the subject of endless concern. He is anxious he should appear a gentleman: 'As in the beginning of them I do add unto my name the title of a gentleman, so I mean to show myself to be one, as well in the outward marks as in the inward man; of the which inward man the music, with the ditties and songs and sonnets therewith joined, shall show to the sufficient judge in that respect' (p. 175).

For a poet who represents himself as dealing so frequently in the covert signs and ambiguous messages of elegant courting, Whythorne has a fear of the uncontrollable interpretation of others. He fears publishing his songs will make him 'a common gaze unto all the world ... [to] hang upon the blasts of all folks' mouths and upon the middle-finger pointings of the unskilful, and also upon the severe judgments of the grave and deep wits' (p. 140). His fascination with his image in the eyes of others is apparent in the numerous occasions on which he has his portrait painted. A mirror, he explains, will not do instead of a por-trait for it shows only a reversed and temporary image, 'for so soon as he looketh off from the glass he forgetteth the disposition and grace of his face' (p. 115). The unstable image of the mirror fails to frame the self as a moral and aesthetic object in the world. What the gazer in a mirror sees is something that is true from his own perspective only; what Why-thorne wishes to see is an objective self as true and graceful to all gazers as to himself: a self-made gentleman, an insider in the private world of courtly dalliance, and a man of moral transparency.

The manuscript autobiography in which he explains the occasions of all his carefully preserved verses, from the merest graffiti scribbled on a wall in Rome to his own version of Psalm 86 presented to Archbishop Parker, may have been begun as an attempt to control these outward written marks of the inward man and fix, for his own satisfaction at least, their exact references. Underlying his contextualising of his verse is a deep anxiety about the anarchic power of language to subvert good order and escape authorial control: 'Before that words be spoken, the speaker of them hath them at his one will to rule; but if they be once spoken before witnesses, then is he no more the master of them' (p. 50).

While Whythorne tries to shape his life to conform to a series of social and moral clichés, the detail of his manuscript creates its own fresher and more eloquent narrative. When, for example, Whythorne amplifies the 'old proverb ... which saith, "This world is but a scaffold for us to play our comedies and tragedies upon"'(p. 175) into a 'sonnet' of a tedious and clumsy predictability, it scarcely catches our attention. But the narrative his autobiography provides of the events to which the sonnet refers, lend the 'old proverb' moving force. He tells of his despair and considerations of suicide when, in his early forties, he finds himself yet again without employment after the double humiliations of a jilting and an unbearable domestic appointment which he is forced to quit (pp. 172–3). After years of attempting to establish himself comfortably in the world, maritally and professionally, he still has only a bachelor's room in London. To develop the theme of man 'wrapped in calamities ... as one who is the receptacle of all worldly troubles and perturbations' (p. 174) in this context is to give the old truisms a newly expressive force. Whythorne's life story of social and economic marginality finds at such moments as eloquent expression as the narrative of professional failure and self-mockery in 'Gascoigne's Woodmanship' or Whitney's tragicomic 'Wyll and Testament'.

Out of a poetry of social aspiration and aphoristic conformity the mid-century writers, almost inadvertently, produced highly usable poetry for a middling sort of educated, ambitious, but economically insecure writers. Imitation of Tottel gave them forms and models with which to articulate their cultivation, their refinement, their inner integrity, and their professional skills. At the same time their poetry registered the social insecurities and marginality of their condition. The self constructed through such poetry could clearly offer a thoroughly marketable, but also an emotionally assuaging, *dramatis persona*.

Notes

1 For 'identity', see OED 2a; for discussion of such terms as 'individual' and 'individuality', see Anne Ferry, *The 'Inward' Language. Sonnets of Wyatt, Sidney, Shakespeare, Donne* (Chicago, 1983) pp. 33–5; for 'autobiography', see Paul Delany, *British Autobiography in the Seventeenth Century* (London, 1969) p. 1.

2 Jacob Burckhardt, *The Civilisation of the Renaissance in Italy*, transl. S.G.C. Middlemore (Oxford, 1945) p. 81.

3 See, for example, David Aers, 'A Whisper in the Ear of Early Modernists; or, Reflections on Literary Critics Writing the "History of the Subject"', in *Culture and History 1350–1600: Essays on English Communities, Identities and Writing*, ed. David Aers (New York and London, 1992) pp. 177–202, on the arbitrariness of the medieval/Renaissance divide; Charles Taylor, *Sources of the Self: The Making of the Modern Identity* (Cambridge, 1989) especially pp. 131–2, for a history of the developing sense of the self in the West; and Jonathan Dollimore, *Radical Tragedy: Religion, Ideology and Power in the Drama of Shakespeare and his Contemporaries* (New York and London, 1984) esp. p. xxvii and chs. 10 and 16.

4 For an essay which studies the influence of the mid-century miscellanies on Spenser and Sidney, and mentions their use of autobiography, see Germaine Wartenkin, 'The Meeting of the Muses: Sidney and the Mid-Tudor Poets', in *Sir Philip Sidney and the Interpretation of Renaissance Culture. The Poet in His Time and Ours*, ed. Gary F. Waller and Michael D. Moore (London, 1984) pp. 17–33. For an earlier study of autobiographical poetry in this period, see Rudolph Gottfried, 'Autobiography and Art: An Elizabethan Borderland', in *Literary Criticism and Historical Understanding. Selected Papers from the English Institute*, ed. Phillip Damon (New York, 1967) pp. 109–34.

5 For a study of what she calls the 'autobiographical assumption' in troubadour lyrics, see Sarah Kay, *Subjectivity in Troubadour Poetry* (Cambridge, 1990).

6 See Arthur Marotti, *Manuscript, Print and the English Renaissance Lyric* (Ithaca, 1995) pp. 135–47. For the increasing privacy of reading habits in the sixteenth century, see Roger Chartier, 'The Practical Impact of Writing', in

A History of Private Life, ed. Philippe Ariès and Georges Duby, vol. III, *Passions of the Renaissance*, ed. Roger Chartier, transl. Arthur Goldhammer (Cambridge, Mass., 1989) pp. 124–5. For the social performance of Tudor courtly verse, see John Stevens, *Music and Poetry in the Early Tudor Court* (London, 1961).

7   For Petrarch, see Ivy L. Mumford, 'Petrarchism and Italian Music at the Court of Henry VIII', *Italian Studies* 26 (1971) 49–67, and for English Tudor verse, see Stevens, *Music and Poetry* pp. 127–32.

8   Whythorne's manuscript was edited using Whythorne's idiosyncratic orthography, by James M. Osborn as *The Autobiography of Thomas Whythorne* (Oxford, 1961). The following year Osborn edited a modern spelling edition with the same title, (London, 1962). When quoting from the *Autobiography* in this essay I shall use the modern spelling edition.

9   *Tottel's Miscellany (1557–1587)*, ed. Hyder Edward Rollins, 2 vols (Cambridge, Mass., 1928–9) II. 7–36. Quotations in my text will be followed in parentheses by volume and page references to this edition.

10   See ibid. II. 99–100, and Marotti, *Manuscript, Print*, p. 144.

11   Wendy Wall, *The Imprint of Gender. Authorship and Publication in the Renaissance* (Ithaca, 1993) p. 97. See also Marotti, *Manuscript, Print*, pp. 215–17.

12   Daniel Javitch, 'The Impure Motives of Elizabethan Poetry', *Genre* 15 (1982) 225–8 (p. 225).

13   *Autobiography*, p. 140.

14   See Rollins (ed.) II. 67.

15   See, for example, Rollins (ed.) I. nos. 2, 12, 26, 265.

16   Ibid. I. nos. 27, 28, 118, 124, 170, 191,194, 200, 295.

17   Wartenkin, 'The Meeting of the Muses', p. 25, notes more generally of the 'civic' voice of mid-Tudor miscellanies, that the poet 'himself stands as a historical reference point behind the experience he relates'.

18   See, for example, ll.86 and 100, and the headnote in *Sir Thomas Wyatt. The Complete Poems*, ed. R.A. Rebholz (Harmondsworth, 1978) p. 439.

19   See Rollins (ed.) II. pp. 74–5 and the preceding discussion. Also Henry Howard, Earl of Surrey, *Poems*, ed. Emrys Jones (Oxford, 1964) pp. 108–9.

20   Rollins (ed.) I. no. 14, and II.141.

21   Thomas Nashe, *The Unfortunate Traveller and Other Works*, ed. J.B. Steane (Harmondsworth, 1972) for example pp. 299, 307, 315.

22   Ibid., p. 316. For the contemporary taste for tournament *imprese* which Nashe mocks, see Alan Young, *Tudor and Jacobean Tournaments* (London, 1987) ch. 5.

23   Michael Drayton, *The Works*, ed. William Hebel, Kathleen Tillotson and Bernard H. Newdigate, 5 vols (Oxford, 1931–41) II. 283.

24   Ibid. II. 287 ('Tagus, farewell' was 'done by the said Earle, or Sir Francis Brian').

25   *The Poems of Thomas Howell (1568–1581)*, ed. Alexander B. Grosart (printed for subscribers, 1879) pp. 108–9. References in parentheses after quotations from Howell will be to page numbers in this edition.

26   Ibid., p. 111. On the well-known gentlemanly reluctance to print, see J.W. Saunders, 'The Stigma of Print: A Note on the Social Bases of Tudor Poetry', *Essays in Criticism*, 1 (1951) 139–64, but see the important modification of Saunders' argument in Nita Krevans, 'Print and the Tudor Poets', in *Reconsidering the Renaissance*, ed. Mario A. Di Cesare (Binghampton, NY, 1992), pp. 301–13.

27 For example, Rollins (ed.), I. nos. 4 or 18 (by Surrey) or 104 (by Wyatt). A later complaint in poulter's measure in *Newe Sonets* has 'To the tune of Winter's Just Returne' (i.e. Surrey's poem; no.18 in Tottel).

28 Michael Brennan, *Literary Patronage in the English Renaissance: The Pembroke Family* (London, 1988) p. 74.

29 See John Keeper's description of this 'abhorde' life in 'The Unsertaintie of seruice' (pp. 153–8).

30 See Barnabe Googe, *Eclogues, Epitaphs, and Sonnets*, ed. Judith M. Kennedy (Toronto, 1989) poems 43–5, and George Turberville, *Epitaphes, Epigrams, Songs and Sonets (1567) And Epitaphes and Sonnettes (1576)*. Facsimile reproduction with an introduction by Richard J. Panofsky (Delmar, NY, 1977) pp. 348–74.

31 The poems were printed in *Epitaphes and Sonnettes* dated to either 1574 or 1576.

32 Whitney may have taken the idea for a last 'Wyll and Testament' from Gascoigne's 'His last wil and Testament' in the Dan Bartholmew of Bathe sequence.

33 Whitney had published poems of courtly dalliance in an earlier volume, *The Copy of A Letter* (1567).

34 See the introduction by Richard J. Panofsky to *The Floures of Philosophie (1572) by Hugh Plat, A Sweet Nosgay (1573) and The Copy of a Letter (1567) by Isabella Whitney* (Delmar, NY, 1982).

35 Quotations from this poem are from the modernised edition in *Women Writers in Renaissance England*, ed. Randall Martin (London, 1997) pp. 289–302 (p. 289). Line numbers in parentheses following quotations are from this edition.

36 Quotations are from the modern spelling version of *The Autobiography* (see note 8 above). Page references in parentheses in my text are to this edition.

37 Osborne discusses evidence for dating in the 1961 edition, p. lxii.

38 Osborne makes this point in the 1961 edition, p. liv. See also two articles by David R. Shore, 'The *Autobiography* of Thomas Whythorne: An Early Elizabethan Context for Poetry', *Renaissance and Reformation*, 17 (1981) 72–86, and 'Whythorne's *Autobiography* and the Genesis of Gascoigne's *Master F.J.*', *Journal of Medieval and Renaissance Studies*, 12 (1982) 159–78.

39 Although an intimate document and never published by the author, the manuscript has a prefatory poem addressed to 'Ye youthful imps', The manuscript begins by addressing 'My good friend' (p. 2).

5
'So Much Worth': Autobiographical Narratives in the Work of Lady Mary Wroth

Marion Wynne-Davies

The initial site of this inquiry will be one of the most readily accepted and easily recognised forms of autobiographical writing, that is, a diary, or more specifically, Lady Anne Clifford's account of her activities in August 1617. In common with other English noblewomen of the early modern period, Clifford paid visits to and was visited by other women of her social group, and during the summer of 1617 she made several calls at Penshurst Place, the home of the Sidney family, which were duly recorded in her diary. For example, on the first of the month she visited 'Lady Wroth' and was joined there by 'Lady Rich'; on the 12th and 13th she 'spent most of the time in playing Glecko & hearing Moll Neville reading the *Arcadia*'; and then on the 19th,

> I went to Penshurst on Horseback to my Lord Lisle where I found Lady Dorothy Sidney, my Lady Manners, with whom I had much talk, & my Lord Norris, she and I being very kind.
>
> There was Lady Wroth who told me a great deal of news from beyond the sea, so we came home at night, my Coz. Barbara Sidney bringing me a good part of the way.[1]

The reason for focusing on these extracts from Anne Clifford's diary will, I hope, be immediately apparent, for they refer to Lady Mary Wroth whose autobiographical writing is the central concern of this essay. However, it would be wrong to neglect the other women who appear in Clifford's diary, since together they form a familial group which was as influential on Wroth's literary canon as on Clifford's diary entry. Indeed, it was precisely these configurations of female kinship

76

which provided the basis of Wroth's construction of her own gendered identity and which enabled her to produce a succinctly feminine version of the Sidneian literary discourse in her autobiographical writings.

One of the most interesting aspects of the group of ladies described by Clifford as visiting one another during this brief period is that they were all, in some way, related to one another, although such is the complexity of their kinship that a family tree is necessary for us to understand the links (see Figure 1). Of course, no such *clef* would have been necessary for the group themselves who, in addition to the period's emphasis upon lineage and descent, would have possessed, as do all familial groups, a mental record of events and allegiances. However, for the benefit of those who exist at a temporal and social distance from the early modern Sidney family, an explanation of the connections represented in the family tree is necessary. To begin, Anne Clifford was the daughter of Margaret Russell whose sister, Anne, was the sister-in-law of Mary Dudley Sidney, Wroth's grandmother. Moreover, Anne Clifford was to become the unhappy second wife of Wroth's first cousin, Philip Herbert. The 'Lady Rich' who visited Clifford and Wroth was most probably Isabella Rich, the daughter of the infamous Penelope Rich who had been allegorised as 'Stella' by Philip Sidney in his sonnet sequence. There was also an association between the two families through the marriage of Robert Dudley (Mary Dudley Sidney's brother) to Lettice Knollys, thereby linking the family to Lettice's children, Penelope Devereux Rich, her sister Dorothy Devereux Percy, and their brother, the even more infamous, Robert Devereux, Earl of Essex. Moreover, despite the fact that Penelope Rich had not married Philip Sidney, she remained a good friend of the family and in 1595 she acted as the young Robert Sidney's godmother. During the subsequent visit of 19 August to the home of Robert Sidney ('Lord Lisle') and his wife Barbara Gamage Sidney, Clifford met three of the second-generation cousins: Robert's daughter ('Lady Wroth'), his daughter-in-law ('Dorothy Sidney'), who was married to the younger Robert and was herself the daughter of Dorothy Devereux Percy and so connected to the Sidneys through her mother, and Philip Sidney's daughter, Elizabeth Manners (the most likely identification for 'Lady Manners'). Completing the group were the Norris family, since it is likely that 'Lord Norris', refers both to the husband, Francis Norris, and his wife, Lady Bridget Vere Norris. This double representation is the most probable interpretation since it would match Clifford's other gender conflations and would also make sense of her subsequent statement, '*she* and I being very kind' (italics mine). Lady Norris was, not unexpectedly, also connected to the Sidneys, since

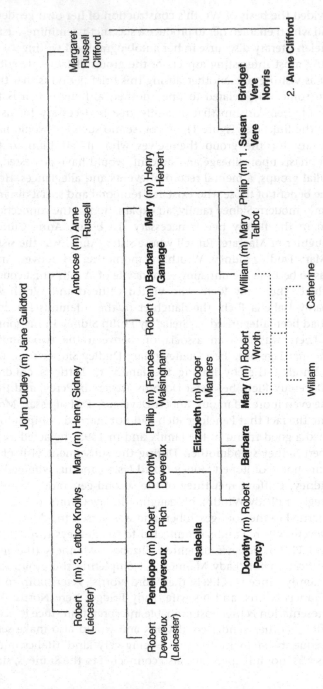

Figure 1 Sidney Women Familial Associations.

she was the sister of Susan Vere Herbert, the Countess of Montgomery, the first wife of Philip Herbert, and it was Susan to whom Wroth dedicated her prose romance, *The Countess of Montgomery's Urania*, and who is represented in her play, *Love's Victory*, as Simeana.[2] Finally, Clifford when at home was forging another familial bond by reading Philip Sidney's prose romance, *The Arcadia*, which had been reworked by his sister, Mary Sidney Herbert, who was of course a well-respected author in her own right. In this way Anne Clifford uses her diary as a record which situates her at the heart of a familial group, enabling her to construct a sense of self which was cultured, informed and supported by her female relatives.

At first it might appear a fortuitous coincidence that another early modern woman writer, Mary Wroth, was present at these gatherings, and doubly fortunate that she should also choose to use this material in her own autobiographical works. Indeed, had more writings from the Sidney family remained extant it might have been possible to add yet another version of events, since Elizabeth Manners was also a respected author.[3] Such a series of 'chances', however, is too far-fetched, particularly since it is becoming increasingly apparent that sixteenth- and seventeenth-century women writers were encouraged by familial groups such as the Sidneys, who counted a number of authors, both male and female, in their ranks. Perhaps, therefore, in such a coterie it would have been surprising if the women attending had not chosen to express themselves in some form of cultural discourse. Regardless of such hypotheses, what is clearly apparent from such multiple uses of these gatherings in textual material is that the supportive system of female kinship served to liberate the writings of early modern women. Mary Wroth therefore, like Anne Clifford, chose to reproduce the female assemblage at Penshurst (although she also refers to the other family homes at Wilton and Baynards) in her play, poetry and prose narratives. Thus, while her writings may not be clearly categorised as 'autobiography' in the straightforward manner of Anne Clifford's diaries, her literary texts are so replete with familial allegory and multiple reworkings of her own life that they demand to be classified within an autobiographical framework.

Before moving on to the manner in which Wroth depicts herself, however, I should like to look briefly at the correspondences between Wroth's and Clifford's treatments of the female community based at Penshurst during August 1617. Penshurst, like all early modern familial houses, functioned as a place where noble women could find pleasure in one another's company without the darker and more dangerous intrigues of the early seventeenth-century court. Indeed, the scene depicted by Clifford presents a strong and supportive female community, and one which, moreover, is decorated with the embellishments of literary texts and references to political intrigues and foreign affairs. Indeed, if we begin to unpick the terms of reference behind Clifford's phrases – 'much talk . . . very kind . . . great deal of news . . . bringing me a good part of the way' – a palpable need for a nurturing and sustaining group becomes glaringly apparent. Anne Clifford herself was particularly reliant on such support since she was involved, with the strong encouragement of her mother, in a lengthy battle to confute her father's will and win her rightful inheritance over a male relative, despite the antagonisms of her husband and the king. Clifford, however, consistently refused to yield, and finally triumphed in default when she inherited the titles in 1643. Lady Mary Wroth's position was, however, little better, since in 1617 she was engaged in an adulterous affair with yet another Sidney, her cousin William Herbert, and had two illegitimate children by him during this same period, possibly giving birth at Penshurst. Moreover, almost all the other women present in Penshurst that August had, or were soon to encounter, similar difficulties. In the younger generation, Dorothy Percy Sidney had been clandestinely married the previous year, and Isabella was to be secretly married the following year. Of the older women, Bridget had frequently lived apart from her husband (who was soon to commit suicide with a cross-bow), and even Barbara Gamage Sidney, the 'noble, fruitful [and] chaste' lady of Ben Jonson's celebrated poem, 'To Penshurst', had arranged a match against the Queen's wishes with the now irreproachable Lord de L'Isle.[4] Thus, each of the women mentioned, with the possible exception of Elizabeth Sidney Manners, were, or had been, excluded from the court world and were thus compelled to use Penshurst as a 'safe' house.[5] For them the refuge offered was not merely a country retreat from which they could emerge refreshed, but an absolute necessity without which they would have been ostracised and disgraced. As such, the congregation at Penshurst, with its secret alliances and clandestine affairs, offered a brief moment of security for its female participants, who were thus able to assert, for however brief a moment, that while protected by these familial surroundings, 'None can accuse us, none can us betray.'[6]

This last quotation is taken from a scene in Wroth's play, and it is to her autobiographical account of the Penshurst gatherings that I will now turn. Wroth's *Love's Victory* is the first dramatic comedy to be written by an English woman, and it is also, significantly, the first play which includes several scenes written entirely for women. These female characters consist of a group of shepherdesses who meet to discuss and recount their experiences of love; as Dalina, one of their number comments:

> Now we're alone let everyone confess
> Truly to other what our lucks have been,
> How often liked and loved, and so express
> Our passions past; shall we this sport begin?
> None can accuse us, none can us betray,
> Unless ourselves, our own selves will bewray.
>
> (III.ii.21–6)

Of course, what Dalina means when she claims that they are 'alone' is that no men are present. The women are thus able to reveal their histories to one another without the threat of patriarchal censure; their 'luck' in love and the number of men they have 'liked' may be openly recounted instead of concealed within a code of social expectation, which allowed only for arranged marriages and a single emotional commitment from women. Moreover, Dalina's final words echo the need for a mutual loyalty amongst women, since the female community ('ourselves') can only be betrayed by individual women ('our own selves'). For the women in Penshurst, with their records of illicit sexual and economic affairs, such a secure grouping must have proved attractive. Moreover, as with Clifford's diary account, several of Wroth's characters who are present in the fictional gathering represent the actual Sidney/Herbert women who joined together on that August day.

In a startlingly autobiographical moment for a play, Wroth depicts in *Love's Victory* an approximation of the coterie of women sojourning at Penshurst. Indeed, it is almost possible to draw up a cast list with the characters' names being neatly paired off with the real women.

Play characters	First-generation women	Second-generation women
Musella	Penelope Rich	Mary Wroth
Simeana	Mary Sidney	Susan Herbert
Silvesta	'Lysa'	Lucy Harrington
Climeana	?	?
Dalina	Penelope Rich	Mary Fritton
Phillis		Lady Philip Sidney
Musella's Mother		Barbara Gamage

The central female protagonist of the play, Musella, clearly denotes Wroth herself, although the alternative representation, that of Penelope Rich, also links Musella to Penelope's daughter, Isabella Rich, who is mentioned by Clifford in her diary. Musella's friend, Simeana, may be identified with Mary Sidney, whose edited version of *The Arcadia* Clifford notes that she is reading, while the second-generation association with Susan Herbert recalls that one of the visitors to Penshurst that August was Susan sister, Bridget Norris. Musella's mother may, of course, be linked with Wroth's mother, Barbara Gamage Sidney. Although, Dorothy Percy Sidney is not directly represented in the play, Phillis denotes Wroth's sister – rather than sister-in-law – so that filial bonds may be said to be recognised in the text. Even Anne Clifford is referred to, albeit obliquely, through the difficulties she was encountering with her father's will, since Musella has also suffered from her dead father's wish that she should marry the boorish Rustic, rather than the dashing and charming Philisses. Thus, given the alterations necessitated by the fictional nature of the text, the parallels still hold true. Indeed, if we extend our search for veiled figures across Wroth's canon, Barbara Gamage and Dorothy Percy are clearly referred to in *Urania I*, and identified as such by the late Josephine Roberts in her excellent edition of the first book of the prose romance.[7] Wroth's parents are, of course, described in the characters of the King and Queen of Morea, and her sister-in-law is figured as Meriana, Queen of Macedon. There does not appear to be a named reference for Bridget Vere Norris, but Veralinda's sisters are mentioned and, as Veralinda shadows Susan Vere, Bridget may said to be included in this group.[8] This means that all the women referred to by Clifford as being in Penshurst in August 1617 are represented in Wroth's work, either in the play or in the prose narrative.

Of course, there are certain matters which even a diary entry must leave out, since no inscribed text may be presumed to be entirely personal, private and safe from intruding eyes. A literary text, on the other hand, while acknowledging autobiographical and familial occurrences, is able both to reveal and to obscure such information through its fictional veiling devices. In this sense, while Clifford's account must remain self-edited, Wroth's version of the women's 'lucks' in love is drawn with sufficient detail to intrigue contemporary and subsequent readers alike into an unending guessing game about the identities of her heroines and their lovers. When in the mid-seventeenth century the Earl of Rutland asked Wroth to provide a *clef* for her canon, he spoke for many of the subsequent readers – and editors – of her work, who sincerely wish that Rutland had been more persistent in encouraging his elderly relative to provide such a decoding device. There was no reason, however, why Wroth, who had chosen to couch her personal history within the allegorical frameworks of sonnet sequence, pastoral drama and prose romance, should suddenly decide to pin down her identifications in the fixed form of an historical table. Moreover, her denial to undertake such a task inevitably makes us question the legitimacy and usefulness of the 'cast list' included above. Glancing back, the instability of the columns so easily constructed begins to become apparent. For example, Musella cannot simply be Wroth since she is also Penelope Rich; Philisses is both Herbert and Philip Sidney; and Simeana may be linked simultaneously to Susan Herbert and Mary Sidney. Identifications slip and change even as they are located, so that any firm 'self' is impossible to locate. Indeed, as Wroth herself points out in Dalina's speech, it is 'our own *selves*' [italics mine] that will be betrayed – not our own *self*. Thus the play is replete with the attendant ambiguities of a series of divided selves, which are set against the more expected and clear-cut identities of the individual characters.

These divided 'selves' are most apparent in Wroth's central characters, Pamphilia/Musella and Amphilanthus/Philisses. The most obvious mirroring, that of Wroth's own love relationship with William Herbert, cannot be missed. Moreover, this is especially true of *Love's Victory*, since the play makes direct reference to the real lovers through the use of puns, a common Sidneian device, initiated by Philip Sidney in *Astrophil and Stella* and reactivated by Wroth in her own works. For example, when Musella defends herself to Silvesta, another shepherdess, against the charge that she has been cruel to Philisses, *Will*iam Herbert is alluded to: 'from poor me he *will* not take relief' (III.i.41–2; italics mine), and when Musella has convinced Silvesta of her sincerity, her friend responds, 'I do believe it, for in so much *worth*, / As lives in you, virtue must needs spring forth' (III. i. 95–6; italics mine), thereby reminding the reader/audience of the author's name as it was commonly reworked during the period.[9] In addition to this relatively straightforward allegory, the songs in *Love's Victory* often directly invoke and answer Herbert's poems, thereby indicating a second level of interpretation, this time literary allegory, although still bounded by the familial referents. For example, at the beginning of Act II the shepherds and shepherdesses draw fortunes from a book which are rehearsed in rhyme, at the end of which Philisses concludes:

> Love and Reason once at war,
> Jove came down to end the jar.

> 'Cupid,' said Love, 'must have place';
> Reason, that it was his grace.
> Jove then brought it to this end:
> Reason should on Love attend;
> Love takes Reason for his guide,
> Reason cannot from Love slide.
> This agreed, they pleased did part,
> Reason ruling Cupid's dart.
> So as sure Love cannot miss,
> Since that Reason ruler is.
>   (II.i.213–24)

These lines are a direct refutation of Herbert's poetical defence of Love against Reason in 'It is enough, a Master you grant *Love*':

> It is enough, a Master you grant *Love*
> At one weapon, 'twas all I sought to prove:
> For worth, not weakness, makes him use but one;
> While that subdues all strength, all Art alone.[10]

Apart from offering a counter-argument to Wroth's uniting of Love and Reason, Herbert neatly puns upon his lady's name, so that it is becomes his adoration of 'worth'/Wroth that makes him a slave to Love and therefore able to refute her argument; in other words, Herbert suggests, she has only herself to blame. This textual bantering between Wroth and Herbert attained its most extreme form when she included one of his poems within her own prose romance and attributed it to Herbert's fictional other, Amphilanthus.[11] However, Wroth's song in the later, and more mature, *Love's Victory* moves beyond this poetic love-play with Herbert, by adding echoes of the contest between Reason and Passion in the Second Eclogues of her uncle's *Old Arcadia* where Reason argues 'that Reason govern most' and Passion replies, in the same comic tone used by Wroth, that 'Passion rule the roast'.[12] This poetic allusion uncovers another identity for Philisses within the familial allegory, for the anagramatised name clearly points towards *Phili*p *Si*dney and not William Herbert. Moreover, this association engenders a domino-like effect revealing further familial identities within the play, for Philisses' sister is called Simeana, a reference to *Mary Sidney*, while the name of his lover, Musella, begins to sound remarkably like *Stella*, in other words, Penelope Rich. Indeed, Wroth self-consciously alludes to this bi-play of classification when she makes Musella ask,

> And yet my true love crossed,
> Neglected for base gain, and all *worth* lost
> For *riches*?
>
> (V.i.3–5; emphases mine)

Here 'riches' suggests Penelope Rich at the same time that 'worth' refers to Wroth. In this manner, any single identification uncovers further 'selves' through a process of intertwining and unending familial allegory, the generational layers perpetuating this decoding process through time as well as place.

However, there is yet another twist to the familial referents. If placed in the context of William Herbert's collected verse it becomes apparent that his poem, 'It is enough, a Master you grant *Love*', defending love against reason was written primarily in response to a companion piece by his friend, Benjamin Rudyerd, which avows the power of reason. Of course, it is highly likely, given the compositional proximity of all the Sidney authors, that Herbert, like his famous uncle and like his prolific cousin, would open his text to multiple identifications. Thus, a double addressee – Rudyerd and Wroth – is perfectly probable. Indeed, it would be unfair not to consider Philip Sidney as a further contributing factor to Herbert's verse, as he so clearly is to Wroth's, even though the linguistic parallels are not so obvious. If, however, Herbert uses his poem to respond to his friend (Rudyerd) and his mistress (Wroth), while simultaneously invoking his famous uncle's (Sidney's) literary inheritance, Wroth can, as always, go one better. Lady Mary Wroth picks up the allusion to Rudyerd and allegorises him in her own texts: in *Urania* he is Ollorandus, Prince of Bohemia and a friend to Amphilanthus (Herbert), while in *Love's Victory* he may be linked to Lissius, the companion of Philisses (again Herbert).[13] In her play Wroth alludes to the poetic debate between Herbert and Rudyerd, following Philisses' lament on the power of love with Lissius's ironic response, 'It seems he [Cupid] missed before he had this guide!' (II.i.225). Thus, Wroth reworks all aspects of the poetic debate: Herbert's argument poem is alluded to in a framework which allows for the discussion with Rudyerd, but which succinctly privileges the response to Wroth herself; she takes the literary acknowledgement of Philip Sidney's verse and rewrites it so that it both emphasises and wittily comments upon the original text and Herbert's rewriting of it; and finally, she couches the Love/Reason dialogue within the primary discourse of Love in all its varying forms thereby ensuring that the 'male' focus upon Reason becomes an insignificant aspect of the whole 'female' concept of Love. On top of all this it is perfectly possible to 'read' Lissius, not as William Herbert's friend Rudyerd, but as Philip Herbert, William's brother and Susan's husband, or, as Matthew Lister, a doctor with whom Mary Sidney has been romantically linked.[14] Again, the female associations outnumber the male links. In this way, Wroth's complex negotiations through a series of autobiographical strands allows her to focus upon her own concerns, to represent her family in a constantly shifting panoply of identifications, to rework her familial literary inheritance, and finally to promote and foreground a 'female' rather than 'male' interpretation on Sidneian genre, theme and character. There is indeed 'so much worth' in Lady Mary's canon, not just in the multiple interpretations referred to above, but even in the repeated and diverse identifications of the author herself.

What the fictional text can never provide, however, is the temporal and locational surety of a diary such as Anne Clifford's. For example, we are left to ponder possible relationships between the time and place of the initial Love/Reason debate between Herbert and Rudyerd, their own inscription of it within the companionate poems, and Wroth's use of the same material (either pre- or post-poetic inscription) in her canon. Thus, while the naming puzzles are an important aspect of Wroth's dramatic strategy (indeed she includes an actual 'riddling' sequence in the play), we must extend the web of literary allusion beyond the walls of individual identification to a system a cultural, philosophical and political beliefs. Wroth's texts can no more be contained within a fixed set of apellative allusions, than any single character in her play, romance or sonnet sequence can be assigned to a single member of the Sidney family. Any autobiographical piece represents a point of convergence between family, time and place. Indeed, if we recall the 'site' of this essay's initial exploration, it becomes apparent that location and temporality are important elements in Lady Anne Clifford's diaries. But, while Clifford had ample reason to be interested in the possession of estates, she is not alone, either in terms of her gender or period, in setting place alongside time as a way of ordering and recording personal information. In parallel, although Wroth is not writing a diary, she echoes the more formal framework of a diary by focusing upon place as the material manifestation of a personal and familial history. Thus, with a final labyrinthine twist, the Sidneian ideology of place becomes inextricably linked to Wroth's creation of a gendered and familial self through the locational device of pastoralism.

In Jan Kip's 1723 engraving of Penshurst Place the physical details of the Sidneian estate are distinct: the house, its formal gardens, the wooded mount rising behind, and the tree-lined walks are all depicted. However,

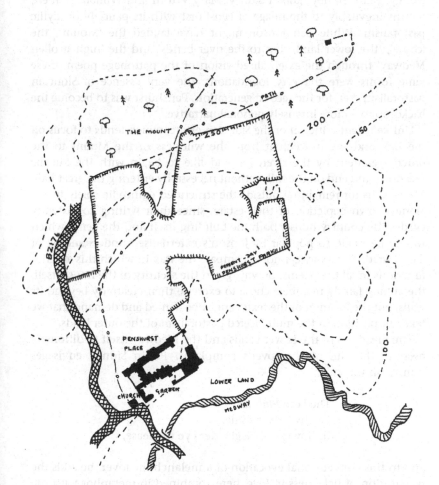

Figure 2

I should like to add to this overall vista, the information that the river Medway runs in front of the formal gardens, from which it is separated by the meadows of the Lower Land (see Figure 2). It is precisely this setting which occurs in Ben Jonson's laudation, 'Penshurst', and which recurs throughout the Sidneian canon. Each authorial voice, Philip Sidney, Mary Sidney, Robert Sidney, Mary Wroth and William Herbert, returns inevitably to the image of Penshurst with its panoply of idyllic pastoralism. While Ben Jonson might have lauded the 'Mount', the 'copse', 'the lower land, that to the river bends' and the 'high-swollen Medway' through the externalised vision of the patronage poem, these same forms were the very foundations, the very essence of Sidneian pastoralism. Yet, for the second generation, Penshurst was to become the backdrop to a more intensely personal narrative.[15]

Unlike recent criticism of the Sidneian estate, which tends to focus on the hall and the movement from the wildness of the Mount to the ordered gardens by the river, I would like to begin with the bucolic meadows and end on the hill, without necessarily entering the great hall. My reasons for remaining outside the structured edifice lie solely in the Sidneys' own depiction of their estate, since their writing consistently evades the country house, both the building itself and the genre which was to be made so popular by Jonson's externalised understanding of 'Penshurst'. This essay cannot encompass the way in which this absence functions in all the poems, or why, given the security of the house itself, the Sidney family members chose to exercise their creativity beyond its walls, but the location of the texts cannot be denied and demand that we take our position in the multifaceted pastoralism of the outer lands.

The meadows of the Lower Lands and the Medway itself traditionally become the site of the lover's complaint. Philip Sidney envisages himself in this role:

> The ladd *Philisides*
> Lay by a river's side,
> In flowry field a gladder eye to please,

and to this conventional evocation of a melancholic lover, he adds the recognition of fickleness in love, here combined in metaphor with the Medway's preponderance to flood: 'you wanton brooke, / So may your sliding race / Shunn lothed-loving bankes'.[16] This same image is echoed in very different poetic situations by the other Sidneys: by Robert in his pastoral on unrequited love and the cruelty of women, 'To drown the fields the angry brooks do move / Their streams'; by Mary in her psalm translations, 'For all the floods'; by Mary Wroth in her representation of abandonment, 'Dangerous fluds your sweetest banks t'orerunn'; and by William Herbert in a depiction of his own lack of steadfastness, 'this wandring will / ... from the natural course stand still'.[17] As with the multiple identifications of the fictional characters, so this peculiarly Sidneian space becomes open to a number of divergent readings. Thus, the Medway becomes a symbol of fickleness in love, both male and female, of uncontrollable anger or passion, and of the overwhelming power of God. In many ways a flooding river lends itself to such fluctuation and mutability of interpretation, but it is a particularly Sidneian technique to take the fixed referents of autobiographical writing and turn them into a series of refracted sites or 'selves', transforming the veiling apparatus of allegory into a series of successive (un)coverings. Indeed, the Sidney/Herbert/Wroth authors evoke similar multiple images of all the key locations on the Penshurst estate. The lower lands, the woods, the walks and the mount are all described in a string of varying pastoral figures: 'Rocks, woods, hills, caves, dales, meads, brooks, answer me' (Philip), 'Meadows, fields, forests, hills' (Robert), 'The woods, the hills, the rivers shall resound / The mournful accent of my sorrowes ground' (Mary Sidney), 'Meadows, paths, grass, flowers, / Walks, birds, brook: truly find / All prove but as vain showers' (Mary Wroth), and 'Looking on the Mead and Grove' (William Herbert).[18] However, as this essay draws towards its conclusions, I wish to turn to a single potent image from these groups of images and focus upon the mount, since it is this site which occupies the key space in the autobiographical writings of Mary Wroth and William Herbert.

In *Urania I* Wroth makes a clear reference to the Mount in her description of the island of Ciprus, where a palace is described as being on a hill, which is surrounded by gardens, an orchard, and 'a fine and stately Wood', while 'at the foote of this Hill ranne a pleasant and sweetly passing river, over which was a bridge'.[19] A similar reference to Penshurst Mount is made, when it figures as the site where Amphilanthus sits alone writing a love poem to Pamphilia: 'in the mid'st of the Wood was a Mount cast up by nature, and more delicate then Art could have fram'd it'.[20] Not unexpectedly, one of William Herbert's poems carries a very similar description:

> ... Let me call t'account
> Thy pleasant Garden, and that leavy Mount,
> Whose top is with an open Arbour crown'd.
> (7–9)

but in Herbert's version the lover is not sitting alone writing melancholy verses to his mistress, but rather asks her,

> Dost thou remember (O securest beauty)
> Where of thy own free motion (more then duty)
> And unrequir'd, thou solemnly did swear,
> (Of which avenging heav'n can witness bear)
> That from the time thou gav'st thy spoils to me,
> Thou wouldst maintain a spotless chastity,
> And unprophan'd by any second hand,
> From sport and Loves delight removed stand,
> Till I (whose absence seemingly was mourn'd)
> Should from a forreign Kingdom be return'd.
>   (10–19)[21]

William Herbert's poetic second-self forges a narrative in which the Mount becomes the site of an amorous encounter, where the lady, Wroth, has given herself to him of her 'own free motion' and promises to remain faithful, even though in the poem she subsequently chooses a different lover (perhaps an allusion to Robert Wroth). The sexual character of their meeting may be uncovered through the phrase '*unprophan'd* by any *second* hand', implying that Wroth's 'spotless chastity' has certainly been 'prophaned by a first hand'. Penshurst Mount becomes, for Herbert, an erotic location in which the pastoral setting evokes a succinctly post-lapsarian Eden, an image hardly recognisable from the chaste melancholy encoded in his lady's verse. The difference is, of course, perfectly explicable in gender terms, since Herbert was heir to a male literary discourse which revelled in sexual allusion; indeed, Ben Jonson appears to invoke a similar characterisation for Penshurst:

> Thou hast thy walks for health as well as sport:
> Thy Mount, to which the Dryads do resort,
> Where Pan and Bacchus their high feasts have made,
> Beneath the broad beech, and the chestnut shade.[22]

Such licence was simply not an option for a noble lady, or indeed any woman wishing to retain her reputation, however besmirched, in the early modern period, and Wroth's depiction of Herbert (in the guise of Amphilanthus) alone on Penshurst Mount, writing forlorn love poems to her, was a far more desirable picture than the cynical and fully requited lover of his own verse.

It is, of course, enticing to imagine that Herbert's more explicit interpretation of familial pastoralism allows us to discover exactly when and where the cousins consummated their sexual passion for one another, to hope that the lovers have indeed 'betrayed' themselves in their texts. It would not be surprising if this particular piece of family gossip is what the Earl of Rutland hoped to learn when he asked Wroth for an explanatory 'key' to her works. But such locational, temporal and personal identifications are as shifting and unstable as the banks of the Medway. For Wroth, as for the other Sidneian authors, the preservation of the 'self' lay in precisely this form of continued obfuscation, of multiple and mutating allegory, of a simultaneous denial and herald-ing of an autobiographical and familial history. To 'confess / Truly', as demanded by Dalina, would indeed be a 'bewray[al]', uncovering both the personal details of the many 'selves' represented by the women of Penshurst, as well as revealing the multiple 'selves' of individuals such as Wroth and Herbert. It is significant that the scene in *Love's Victory* where the shepherdesses choose to divulge their autobiographical nar-ratives is precisely the one in which Musella (Wroth) never appears. She is mentioned only once, when Phillis confesses,

> Nor did I ever let my thoughts be shown
> But to Musella, who all else hath known.
> (III.ii.69–70)

Musella is both absent on stage and yet, through her all-embracing know-ledge, also present. Similarly, Wroth may be simultaneously removed from an immediate presence in her literary canon, her autobiograph-ical narrative both obscured and revealed. It is precisely this evasion of the sureties of Clifford's diary entries that enables Wroth to retain the continued and multiple identifications which literary autobiography makes possible, and moreover, to ensure that while she might have 'confess[ed] / Truly', she never 'betray[ed]' herself or, indeed, any other female member of her family.

Notes

1 *The Diaries of Lady Anne Clifford*, ed. D.J.H. Clifford (Stroud: Alan Sutton, 1990) pp. 60–1.

2 Indeed, Philip Herbert had been betrothed to Bridget before Susan. The connection between the sisters was affirmed by the fact that Bridget's daughter was placed in Susan's household to be brought up.

3 Ben Jonson, 'Conversations with William Drummond', in George Parfitt (ed.), *Ben Jonson. The Complete Poems* (Harmondsworth: Penguin Books, 1975) p. 466.

4 Ben Jonson, 'To Penshurst', pp. 96–7.

5 The idea of a 'safe house' is central to the book I am working on at present, *Safe Houses: Familial Discourse in the English Renaissance*.

6 *Love's Victory* III.ii.25, in *Renaissance Drama by Women: Texts and Documents*, ed. S.P. Cerasano and Marion Wynne-Davies (London: Routledge, 1995) p. 108; all future references to the play will be made parenthetically.

7 Mary Wroth, *The First Part of the Countess of Montgomery's Urania*, ed. Josephine A. Roberts (Binghampton, NY: Medieval and Renaissance Texts and Studies, 1995).

8 For the first two identifications I am indebted to Josephine Roberts, whereas for the two latter characters I have used my own reading of Roberts' edition; see Roberts, (1995) pp. xcii–iii, 107, 190 and 804.

9 Josephine A. Roberts, *The Poems of Lady Mary Wroth* (Baton Rouge: Louisiana State University Press, 1983) p. 18.

10 William Herbert, *Poems Written by the Right Honorable William Earl of Pembroke* (London: 1660) p. 7.

11 Roberts, (1983) p. 44.

12 Philip Sidney, *The Countess of Pembroke's Arcadia (The Old Arcadia)*, ed. Katherine Duncan-Jones (Oxford: Oxford University Press, 1985) p. 119.

13 Roberts, (1995) pp. xci–xcii.

14 Cerasano and Wynne-Davies, pp. 94–5.

15 Jonson, ll.10, 19, 22 and 31.

16 Other Poems 5, ll.1–3 in *The Poems of Sir Philip Sidney*, ed. William A. Ringler (Oxford: Clarendon Press, 1962) pp. 256–9.

17 Robert Sidney, 'Pastoral 8' (*The Poems of Robert Sidney*, ed. P.J. Croft (Oxford: Clarendon Press, 1962) p. 209); Mary Sidney, 'Psalm 50' (*The Triumph of Death and Other Unpublished and Uncollected Poems by Mary*, ed. Gary Waller, *Sidney, Countess of Pembroke* (Salzburg: Salzburg University Press, 1977) p. 106); Wroth from 'Song 4' in *Pamphilia to Amphilanthus* (Roberts (1983) p. 113); and William Herbert 'A Sonnet' (Herbert, p. 41).

18 Philip Sidney in *Sir Philip Sidney. Selected Poems*, ed. Katherine Duncan-Jones (Oxford: Clarendon Press, 1973) p. 101; Robert Sidney, 'Song 3' (Croft, p. 153); Mary Sidney, 'The Dolefull Lay of Clorinda' (Waller (1977), 176–9); Mary Wroth, *Love's Victory* I.ii.17–19 (Cerasano and Wynne-Davies, p. 98); and William Herbert, 'A Sonnet', p. 40.

19 Roberts (1995) pp. 47–8.

20 Roberts (1995) p. 133, and Roberts' notes on p. 731. See also a number of poems in *Pamphilia to Amphilanthus* where love is said to have been first acknowledged in a wood or arbour (see *Women Poets of the Renaissance*, ed. Marion Wynne-Davies (London: Dent, 1998) pp. 183–228.

21 William Herbert, pp. 56–7. Roberts also quotes this poem in relation to Wroth's description, but her transcription appears to be different from my own; Roberts (1995) p. 731.
22 Jonson, ll.9–12.

6

'Child of Time': Bacon's Uses of Self-representation

W.A. Sessions

In 1620 Francis Bacon published his *Instauratio Magna* or the *Great Instauration* (or *Renewal*). Bacon's massive Latin text was to be one more – for him, the climax – of the series of 'recoveries' (the term with which the period viewed these phenomena) that had marked the English Renaissance from the time of Henry VIII – 'recoveries' that were, in point of fact, more often breaks with the past. What is strange is that, in the text of this ostensibly philosophical instauration, Bacon injects a series of self-revelations. More strangely, these subjective discourses may appear staged in the midst of scientific definitions and revolutionary proposals, but their dialectical presence defines, in fact, the narrative or the flow of argument Bacon is presenting. That is, Bacon's objective text proceeds not from a supra-historical logic (whether Pythagorean or Stoic, Thomist or Calvinist) but from a *self* in the midst of its own terrifying history. As though to make this point as clear as possible for his reader, Bacon begins his text with the bold capitals and special printer's type and setting that reveal an author or, more strictly speaking, a narrator almost unbelievably self-centred: 'FRANCIS OF VERULAM / Reasoned Thus With Himself, / And Judged It To Be For The Interest Of The Present and Future Generations That They Should Be Made Acquainted With His Thoughts.' Deliberately echoing Julius Caesar's autobiographical Gallic histories, the Latin is even more solipsistic: 'SIC COGITAVIT; / TALEMQUE APUD SE RATIONEM INSTITUIT / quam viventibus et posteris notam fieri ipsorum interesse putavit.'[1] What is Bacon doing here, taking on not only Julius Caesar's third-person autobiography but the role of self-ordained prophet judging time?

For one thing, he is centralising his text. He is entrapping or seducing his reader, whom he must convert to a new way of thinking. In fact,

Bacon is extending an old strategy of the classical oration. The Baconian difference from the Greek and Roman strategies is, however, that Bacon makes his orator's *ethos* – the staged speaker's self – into a saint-like figure who is giving birth through his suffering to the modern world. 'Being convinced that the human intellect makes it own difficulties', Bacon's first line in regular type begins and then goes on,

> he thought all trial should be made, whether that commerce between the mind of man and the nature of things, which is more precious than anything on earth, or at least than anything that is of the earth, might by any means be restored to its perfect and original condition, or if that may not be, yet reduced to a better condition than that in which it now is... (4:7)

This is not so much historical or philosophical parody as religious. A new rhetorical strategy is at work.

In this opening to the *Instauration Magna*, the *ethos* and its self-figuration lead next to a spectacular passage where Bacon uses self-aggrandisement ironically to demonstrate himself (through a fascinating defence lawyer's logic) as a model of humility. With stunning originality, Bacon places a totally new typology of scientific piety within himself:

> For my own part at least, in obedience to the everlasting love of truth, I have committed myself to the uncertainties and difficulties and solitudes of the ways, and relying on the divine assistance have upheld my mind, both against the shocks and embattled ranks of opinion, and against my own private and inward hesitations and scruples, and against the fogs and clouds of nature, and the phantoms flitting about on every side; in the hope of providing at last for the present and future generations guidance more faithful and secure. Wherein if I have made any progress, the way has been opened to me by no other means than the true and legitimate humiliation of the human spirit.

Unlike predecessors who have not examined 'facts and examples and experience' and for whom 'invention' is 'nothing more than an exercise of thought' or worse, 'to invoke their own spirits to give them oracles, Bacon the humble scientist stands out:

> I, on the contrary, dwelling purely and constantly among the facts of nature, withdraw my intellect from them no further than may suffice

to let the images and rays of natural objects meet in a point, as they do in the sense of vision; whence it follows that the strength and excellency of the wit has but little to do in the matter. And the same humility which I use in inventing I employ likewise in teaching.

There, 'I have not sought (I say) nor do I seek either to force or ensnare men's judgments, but I lead them to things themselves and the concordances of things, that they may see for themselves what they have' for disputation and what they can contribute to. Indeed, referring both to his scientific method and his prose style of aphorisms, 'I so present these things naked and open' that, Bacon adds in more erotic imagery, 'I have established for ever a true and lawful marriage between the empirical and the rational faculty, the unkind and ill-starred divorce and separation of which has thrown into confusion all the affairs of the human family' (4:18–19).

This staged autobiography operates throughout the whole of the *Instauratio Magna*, and at one crucial point Bacon repeats his preface-motif of honesty of representation. This is in his 92nd aphorism of the *Novum Organum*, Bacon's text following his various prefaces in the 1620 *Instauratio* that he intends as an updated version of Aristotle's *Organon*. In the middle of this aphorism, rhetorically and epistemologically climactic in the whole text, Bacon uses the first person as he had not for some time: 'I am now therefore to speak touching hope; especially as I am not a dealer in promises, and wish neither to force nor to ensnare men's judgments, but to lead them by the hand with their good will.' Bacon can compare himself, in fact, to one of the great cultural heroes of his time in a brilliant device of self-justification:

And therefore it is fit that I publish and set forth these conjectures of mine which make hope in this matter reasonable; just as Columbus did, before that wonderful voyage of his across the Atlantic, when he gave the reasons for his conviction that new continents might be discovered besides those which were known before; which reasons, though rejected at first, were afterwards made good by experience, and were the causes and beginnings of great events. (4:91)

Thus, for Bacon, all his personal achievement can be attributed to a new conceptualisation of time, one that he makes personal. Bacon defines himself as a child of time: 'And therefore I attribute my part in all this, as I have often said, rather to good luck than to ability, and account it a birth of time rather than of wit' (4:109). Autobiography – subjectivity –

exists in a special relationship not to intellect but to history, crucially not a dogmatised history but one open to fortune and uncontrolled change. Achievement, the meaning only subjectivity can give reality, has a new centre in a new conception of history.

If viewed, therefore, in the context of this immediate work of 1620, the *Instauratio Magna*, Bacon's staged autobiography can be seen to serve a clear purpose. As the language of his calculated asides reveals, Bacon gives his methodical text – a work he sees as offering nothing less than his solution for what he saw as the chaos of modern Europe – a confessional frame. He invents an Augustinian sincerity that not only provides a voice for the new science and its human dialogic figuration but gives that voice a literary tradition. Modern science has found a voice in these asides such as has never existed before. In fact, Bacon's ideological effect of voice and figuration would be as dynamic as any in the seventeenth century, as both Leibniz and the Royal Society would prove. Here was not a mechanical text being offered in the name of an abstract science, but an actual New Logic, a method of reality imbued with the voice and figuration of a suffering human being. Bacon thus reveals himself, in instances of staged autobiography (recurring far more than I can indicate in this brief study), as the true master of rhetoric and parody. That is, the whole effect Bacon intends with his uses of autobiography springs from a calculated rhetorical ploy that will entrap and convert a reader. Bacon is making a variation – a transformative one – on a Renaissance tradition of subjectivity that he would redefine not only in his preface to the *Instauratio Magna* but more fully in his greatest example of fictional autobiography, the *New Atlantis*. Before examining the roots of this tradition as it came to Bacon, however, and his crucial variations on this humanist tradition, I want to look further at what Bacon intended with this rhetorical ploy of subjectivity in the midst of objective scientific proposals designed to bring peace to the societies of both contemporary England and Europe.

*

If, therefore, it is patently clear that the Jacobean master of rhetoric and legal proof does have a strategy working in his use of autobiography, the question still remains: why these personal outcries, like theatre monologues, amid legal aphorisms whose objectivity of proof would rescue science and learning from the subjective webs of Aristotelian and Scholastic science or from the dogmatic and emotional induction of the ant-like Alchemists by establishing the New Organum or Inductive

Logic with its bee-like harmonies? The answer is immediately simple: it is part of a process by which Bacon would convert the reader to action. As throughout his *Instauratio*, so in these outcries Bacon needs to dramatise his discourse in order to draw his reader into the process of renewing science and learning. To that end, Bacon's *ethos* keeps revealing the author himself – especially his personal voice – as a means of identity for the reader. He would render for his reader the intense subjectivity that has driven this writer – no one less than the Lord Chancellor of the kingdom – to make a new scientific world-picture out of the old. As with the orators of the ancient world and in 1620 with the dramatised 'sincerity' of the preachers of post-Reformation Europe can captivate and convert his readers. If Luther and his cry of 'ICH kann nicht anders' centre the development – perhaps the origination – of autobiography in the modern world, Bacon as the child of that tradition can dramatise now the dilemma of the searching self by dramatising himself. The new scientist thus becomes, in Bacon's new theatre of philosophy, not the heartless logician of the Aristotelian–Thomistic tradition, nor the bizarre and anti-social Alchemist, nor a self-centred Bruno. He acts as Bacon's Father of Solomon's House does in the *New Atlantis*, and as did the earlier figuration of the Father in the wise old teacher in Paris, who sits among his students, in Bacon's *Refutation of Philosophies*. In fact, the Father of the *New Atlantis*, as described by the searching 'I' of Bacon's Utopia, appears like Bacon himself: 'He was a man of middle stature and age, comely of person, and had an aspect as if he pitied men.' Bacon had viewed himself as just such a heroic figure in his earliest staged autobiographical text. In a letter of 1592, at the age of 31, he had written to his uncle, Lord Burghley, of his 'vast contemplative ends' and his heroic dream for himself: 'I have taken all knowledge for my province.' The Father of Solomon's House is Bacon's dream-fulfilment at the end of his life.

But no prefiguration of autobiography could equal the stunning dialectical effect the Lord Chancellor brought to his greatest representation of his system in the *Instauratio Magna* through his figuration of the humble orator/teacher/father. In this 1620 text, which marked the height for all the achievements of Bacon's entire life, the actual statements in the opening discourses of the *Instauratio Magna* pose a narrative dialectic, one of the most skilful in Bacon's canon and indeed in any major English literary work of this period. The dialectic dramatises the figure of Bacon as few earlier major philosophical texts ever had its author, not even Lucretius in his emotional invocation to Venus in *De rerum natura* although, after Bacon, Descartes would frame his *Discourse* with a special Augustinian self-consciousness (but never in such direct appeal to the reader). Thus, in the *Great Instauration*, as elsewhere in passages of the *Advancement of Learning*, not only is subjectivity placed as a corollary to the objective forms of scientific logic that follow in the texts of the *Novum Organum* but it is placed as a necessity, the self-recognition and resolve needed for the task ahead. The converted reader has a model.

The autobiographical passages thus perform a propaedeutic purpose. The reader is invited to take on the identity of a Bacon so dramatised, congratulating the author and performing with him. Reader and author will wind their way together, subjectivity wedded to subjectivity, so that what follows is not the cancerous mechanical insincere logic of the past but the warm, intimate, even erotic logic of the present, reader and author united. Bacon is dramatising here the kind of friendship (the 'ad filios' intention he wrote in his notebook of 1609, *Commentarius Solutus*) that he always sought for his readers and himself, and that he dramatised as a way of reading or entering most of his texts. Together the friends can wander through the labyrinth of the disjunctive aphorisms that themselves model the method of time. With almost legal calculation, then, subjectivity now re-enters the objective social mode of scientific redemption, and the first step – the purifying moment, the cleansing ritual – is the recognition of the prophetic figuration of the lonely Bacon who has brought his texts, at a price, to the reader. Autobiography thus joins, in its intense subjectivity, with the reader and leads this captivated 'friend' to the task of redemption and social renewal that such a text as the *Instauratio Magna* calls for. Like the Lutheran reformation and the Calvinist call that followed it, the 'heart' is transformed in a new system of conversions.

*

But where had such a vogue of subjectivity originated in early modern Europe? The range of sources is infinite, of course, but at least one figure appears to have thought through the concept quite early – Erasmus – and in one spectacular text – his 1501 *Enchiridion militis Christiani* (Handbook of a Christian Soldier). By the time of the *Magna Instauratio*, almost 120 years later, Erasmus' *Enchiridion* – its teaching and its dialogic form – had penetrated to the marrow of European culture through its offspring in More (especially the *Utopia*), in Luther, and then figure after figure in works for wide audiences as varied as *Hamlet* or Burton's *Anatomy of Melancholy* or the autobiography of Sir Thomas Browne, *Religio*

Medici, written in the middle of the English Revolution. Given this influence, it is essential to look, however briefly, at this seminal text and its originating rhetorical structures. Such a look will show not only how Bacon is truly the child of Erasmus (however ontologically different), but also how Bacon uses a central text of early modern Europe (and the culture that sprang from it) for his own redemptive purposes.

In Erasmus' 1501 text, with its special early modern version of the Augustinian mode of reality (first transformed for the Renaissance, of course, by Petrarch's staged autobiography), the Dutch monk announces: 'Be assured there is nothing so true, nothing so certain and beyond all doubt of all that you take in with your ears, behold with your eyes, or touch with your hands than what you read in these writings.'[2] Not only is Erasmus making his own variation on the nominalist dialectic of the English Franciscan monk William of Ockham of a hundred years before, but he is adapting the anti-Thomist perception of reality to a contemporary and thoroughly modern situation. With a special irony, he is as completely soteriological in his intent as any missionary. He is proposing a new form of redemption. Erasmus is writing a 'short way to Christ' (127) for a young German courtier in 1501. The young man has grown weary of the ritualised contemporary court world and is wondering how to 'escape from Egypt, with her sins and allurements' and 'follow Moses on the road to virtue' (24). In such interiority, as Erasmus will demonstrate throughout his text, ceremonies have little validation for the self. The court no longer has epistemological validity but is simply one more object to be used and analysed like any other. It is certainly no centre of social or personal transcendence, as was held to be the Burgundian court of Charles the Bold and his offspring – a court from Erasmus' own world whose *gloire* he specifically loathed. Because of its calculated sincerity, the *Handbook* and 'short way to Christ' soon became one of the fundamental texts for the English Renaissance and Reformation. It translated the ancient Paulist topos of the Christian soldier into a new world-situation, one increasingly pluralistic, a multiplicity Erasmus' methodical terms could encounter and transform better than the dominant method of Scholastic logic. Such an encounter was dependent on the strength of individual subjectivity – the transformed 'heart' – a strength the self needed to survive such a worldly disaster as the European court. Above all, for Erasmus, in the ideological chaos of modern Europe, the self could not depend on external world-systems. Thus, in the Erasmian mode, any new conversion to a new method, scientific or philosophical, for human existence or any system at all designed to transform the world must start with the subjective. It only begins in the 'heart'.

In fact, for Erasmus, the only remedy against 'ignorance' as evident, for example, in such a 'sea-shrine' as he calls it, of Our Lady of Walsingham (in Norfolk just across from Flanders) – an 'ignorance' dominating the culture of the desperate young German courtier – is to read Scripture. It must be read 'with your whole heart, with the deep and unshaken conviction that there is not one tiniest detail contained that does not pertain to your salvation' (55). In fact, in this 1501 new logic of the utter subjectivity of all human reality focused in a liberating text, Erasmus proclaims: 'If you destroy what man made, you will restore what God made' (90). Christ is both Logos-Word and Gospel-Word – combined *within* the newly liberated Christian who reads. It was thus a short step from the 1501 *Enchiridion* to the internalising that could lead to the universal freedom Erasmus announced he sought. 'I wish that every woman might read the Gospel and the Epistles of Saint Paul,' he wrote in his 1516 preface to his Latin translation of the New Testament from the original Greek. In this preface to a text Luther would require immediately for his students in Wittenberg, Erasmus would also add, as further readers and internalisers of new texts, other marginalised human beings like the Scots, the Irish, the Turks and Saracens (in that order!). Erasmus even hopes, in a famous passage that set the pluralism of honour in the modern world as early as any, 'that the farmer might sing snatches of Scripture at his plough' and 'the weaver might hum phrases of Scripture to the tune of his shuttle'.[3] Not surprisingly, More's *Utopia* that year would describe a possible new world where 'in the churches no image of the gods is to be seen, so that each man may be free to form his own image of God after his heart's desire, in any shape he pleases'.[4]

Thus, by 1620, and after the almost totally successful revolution begun by Erasmus in which a new conception of the subjective had emerged in northern Europe, Bacon had to consider the new reading habits of his audience and their culture. Faced with a world in which concepts of subjectivity frame and define the new realities of modern life, Bacon needed to determine how to use this new framing of reality for purposes of presenting his own objective structures of reality, social and historical. He must transform and move from that utter abandonment of self for the greater self of Christ the reformers had taken from Erasmus' *Enchiridion* (a teaching about self ingrained in Bacon both by his culture and by his Calvinist mother). But just how? His problem of the right dialectics for representation of his ideas was greater than a cultural historian might imagine in the twentieth century. Bacon must use the dominant discursive mode of his time for his own calculating purposes – as would those staged autobiographical texts of Bacon's contemporary Donne and Bacon's young friend George Herbert which begin with Petrarch and move on to new ideologies of subjectivity. As Bacon knew well, there could be no hope of winning an audience to support the new endeavours of science and technology that he saw as essential for the redemption of modern European society if he did not adapt a strategy such as he had outlined in his 1609 notebook *Commentarius Solutus*. He was quite specific there: 'To consyder wt opynions are fitt to nourish tanquam Ansae and so to grift the new upon the old, ut religiones solent [as is the custom of religions]' (11:65). This would be essentially a strategy of parody, but a profound parody, as Bacon's reference to the redemptive methods of religion indicates. Parody would be not so much a matter of mockery or even of comparison as of stimulating analogy. Bacon's would be, therefore, a method of profound cultural significance, as Marx himself later recognised. It became, in fact, the strategy the Italian communist Gramsci called for in the twentieth century: to infiltrate a culture with a new system of salvation, not by head-on aggression, but by converting the dominant discourse to the uses of a greater ideological system which would finally transform the culture surrounding the first discourse.

No work exemplifies this strategy of infiltration and accommodation better than Bacon's own Utopia, the *New Atlantis*, published a few months after his death. The narrative is told entirely in the first-person, first from the point of view of the 'we' of a destitute crew washed ashore on an unknown island in the south Pacific, then narrowing at the end to the 'I' who is personally received and catechised by the majestic Father of Solomon's House – a scientific institution like a super Cal Tech or MIT or Salk Institute – which rules the island politically and epistemologically. It is the 'I' who is converted to the wonders of the island at the end through the series of steps that begin in the disaster of shipwreck (the 'zero-degree', in Foucault's phrase, required in Bacon's conceptualising of history). The narrative thus ends in the conversion required for most endings of popular science fiction (especially the Spielbergian kind). Before that closure, however, a process of intense subjectivity in a totally new environment – what Defoe turned into his discursive narrative of *Robinson Crusoe* less than a century later – marks the learning curve of the 'I'. The process of the narrative – subjectivity transforming itself into an objective social identity – leads the literally and culturally shipwrecked European sailor to convert to the new world of Solomon's House and the new Jerusalem of Bensalem. What happens then in the narrative of Bacon's scientific Utopia is precisely what happens in the learning curve, so to speak, of the *Magna Instauratio* and the scientific proposals that Bacon believes can cure the sickness of England and of modern Europe. The reader is engaged in a rhetorical dialectic in which the effects of staged autobiography will transform the most objective discourse into a subjective call for a new scientific logic and a purged society.

Finally, Bacon's great problem of representation was to turn the Erasmian rhetorical mode of subjectivity that dominated the great texts of seventeenth-century England into a mode of action that would stimulate the reader-self to enter society and redeem it from the terrible negativity of the late Renaissance through his new scientific logic. The method was more successful than not. In fact, the dynamics by which Bacon became the cultural icon of Europe for the next two centuries developed in no small part from the success of Bacon's new rhetoric of autobiography and its dialectic. In a Baconian world of labour, in the Virgilian and Lucretian senses he intended, subjectivity exists only in relationship to the Other of society and objectivity. The dialectic of autobiography, certainly as Bacon uses it in his texts, reveals that relationship of self to Other as its deepest meaning, with a certainty that had not existed in Europe since St Thomas Aquinas. With Bacon, autobiography is now verified in the framework of a conversion-system no longer religious and thus open to other cultural forms of experience. In fact, in Bacon's 1620 *Instauratio Magna*, a moment of freedom in the history of Western discourse has been defined in terms quite different from previous representations of systems of salvation. The dialectical terms of this moment are entirely Bacon's, whatever their earlier origins. Their effects are clear and operative throughout Bacon's canon. Subjectivity may become historical in his texts, but historical action is validated only, in Bacon's dialectical concept of labour, through a subjectivity, either author's or reader's, dramatised as authentic and living.

Notes

1 *The Works of Francis Bacon* ed. James Spedding, Douglas Denon Heath and Robert Leslie Ellis (1857–74, Stuttgart-Bad Cannstatt: Friedrich Fromman Verlag Gunther Holzboog, 1963) 4:7; pp. 1–121, hereafter in text.
2 *Handbook*, transl. and annotated Charles Fantazzi, in *Collected Works of Erasmus*, vol. 9, ed. John W. O'Malley (Toronto: University of Toronto Press, 1988) p. 55, hereafter cited in text.

3 Collected Works of Erasmus (Toronto: University of Toronto Press, 1981) pp. 2–146.

4 Thomas More, Utopia, transl. and ed. Robert M. Adams (New York: W.W. Norton & Co., 1975) p. 86.

7
Her Own Life, Her Own Living? Text and Materiality in Seventeenth-century Englishwomen's Autobiographical Writings

Helen Wilcox

The focus of this essay is a selection of texts from the considerable body of autobiographical writing by seventeenth-century Englishwomen, ranging from the diary of Lady Anne Clifford at the beginning of the century to the self-vindication of Anne Wentworth in the Restoration period. My purpose is to investigate the interrelation of the idea of a life in writing with the material realities of living. Autobiography is a genre – or perhaps an instinct – which is always poised at the borderline of word and deed. This balancing-act is rendered additionally complex in women's autobiographical texts by, on the one hand, the gendered sense of a life instilled in their authors and, on the other, the women's limited material independence. The light which these two constraining factors shed on one another, and on the emergence of forms of textual self-representation among early modern Englishwomen, is my concern here.

The autobiographical mode of writing was extremely popular among seventeenth-century Englishwomen, for a range of reasons, which include the Protestant emphasis on devotional self-examination, the association of feminine experience with the domestic or social sphere, and women's limited access to the more classical genres of literature.[1] The flexibility of the form of autobiographical expression – it could be given shape in diaries, letters, confessions, meditations and defences as well as in more formal verse or prose memoirs – made it particularly appealing to women.[2] These varied autobiographical writings contain intriguing evidence of the tensions between textual and material lives. What did Margaret Cavendish mean, for example, when she referred to her autobiography as 'her own life'?[3] And what are the subtle differences, in meaning and in real experience, between that phrase and the more materially-oriented 'her own living' for women in seventeenth-century England?

My first example comes from a fascinating autobiographical poem written in 1632, which gives a very clear image of the pattern of a seventeenth-century woman's life – and living. Martha Moulsworth wrote her 'Memorandum'[4] in her fifties, after the death of her third husband, and the framework of her life as set out in the poem appears utterly patriarchal. She was the daughter of her 'heavenly father' before all else, and was brought up by her 'earthly parents', though her mother remains unmentioned in the poem. Her father is described as a man of 'spotless fame' and 'gentle birth':

> He had, and left lands of his own possession
> He was of Levi's tribe by his profession
> His mother Oxford knowing well his worth
> Arrayed in scarlet robe did send him forth.
> By him I was brought up in godly piety
> In modest cheerfulness, and sad sobriety
> Nor only so, beyond my sex and kind
> He did with learning Latin deck [my] mind
> And why not so? The muses females are
> And therefore of us females take some care
> Two universities we have of men
> Oh that we had but one of women then
> Oh then that would in wit, and tongues surpass
> All art of men that is, or ever was
> But I of Latin have no cause to boast
> For want of use, I long ago it lost
> Had I no other portion to my dower
> I might have stood a virgin to this hour
>
> (lines 23–40)

At this point, Moulsworth adds an ironic note in the margin of the manuscript: 'Latin is not the most marketable marriage metal.' The woman's relation to the 'market' is absolutely clear; it is not her learning or her 'wit' which are to be marketed, but her virgin self, along with an attractive dowry in another kind of 'metal' – coinage. The woman enters the marketplace not as a merchant to sell her creative work, but as produce herself.

Moulsworth's poem (which remained in manuscript until as recently as 1993) suggests by its tone that she was no passive woman; her comments on the ideal of a women's university indicate spirit and vision. However, though independent of mind, she was dependent in material terms upon her father and, subsequently, her husbands. The poem continues by introducing her three husbands as a group:

> But though the virgin muses I love well
> I have long since bid virgin life farewell
> Thrice this right hand did holy wedlock plight
> And thrice this left with pledged ring was dight
> Three husbands me, and I have them enjoyed
> Nor I by them, nor they by me annoyed
> All lovely, loving all, some more, some less
> Though gone their love, and memory I bless.
>
> (lines 41–8)

At this point Moulsworth proceeds to introduce each husband individually, but the preceding lines are worthy of note for their balanced outlook towards the husbands in general, epitomised by the reciprocal syntax of 'I by them' and 'they by me' in line 46. She may well have been 'enjoyed' by the husbands who chose her at the marriage 'market', but, equally, Moulsworth asserts that she 'enjoyed' them. Her honesty in this autobiographical poem is striking; she has no scruples about saying, for instance, that she loved 'some' of her husbands more than others. The subsequent descriptions suggest that the third husband, Bevill Moulsworth, was her favourite: she writes later in the poem that he was 'buxom' to her and allowed her to lead an 'easy darling's life', which interestingly includes having her will 'in house, in purse, in store'. Pleasure in wifehood, it seems, involved having control over the 'purse'; her independence, modest though it may have been, was partly perceived in financial terms.

At the conclusion of the poem, Moulsworth admits that she is wary of 'knitting here a fourth knot' – that is, of marrying again – and the final couplet explains why:

> The virgin's life is gold, as clerks us tell
> The widow's silver, I love silver well.
>
> (lines 109–10)

In this witty conclusion she not only hints at her contentment with the widow's state, but uses the traditional classification of 'gold' and 'silver' to suggest one rather material reason for her pleasure: she loves 'silver' and for once has some in her own pocket. No wonder widows were the butt of many a joke in the jest-books and plays of this period; they were the one category of women who had financial independence. The original source of their 'silver' was still patriarchal, but having passed through the stages of financial dependency – daughter and chattel; virgin with dowry; wife and mother – widows, particularly those without surviving children as in Moulsworth's case, gained a freedom which was often perceived as threateningly unchecked.

Martha Moulsworth, writing with the benefit of over 50 years' experience, could see at a glance in her poetic 'Memorandum' just how her life had been shaped by the marketing not of her Latin skills but her more conventional dowry. Lady Elizabeth Delaval, on the other hand, writing about her teens when her own financial and marital future remained uncertain, expressed in her autobiographical *Meditations* the pressure of anxiety about money. Though conscious of how fortunate she was to be of the courtly class and therefore not spending all her days 'labouring for food and rayment', as she noted in 1663, she nevertheless had to negotiate quite firmly for a modest element of financial identity:

> When all things were concluded betwixt Mr. DeLaval's friends and mine for our maryage, I absolutely refused to consent to it till my aunt Stanhope (in whose hands my thousand pound was left) had first pay'd me that mony to disposs of as I pleas'd.
>
> My father and my aunt Stanhope intended it shou'd have been a part of my portion and did not at all consern themselves with takeing care about my debt's, which I thought a very great hardshipe towards me, since had they not been pay'd before I was a wife, they must certenly have fallen upon my husband, which I might very probable have been many times reproach'd withall by his relation's and have lived for that reason (if for no other) unhapily amongst them.[5]

Delaval had been left the thousand pounds by her grandmother and clung to it as a part of her own self, not a sum to be absorbed into her 'portion' or marriage dowry. Besides, she needed it to settle the debts contracted during her time at court, which she was keen not to carry with her into marriage. Her reasoning is shrewd: a wife who brings debts into a marriage will be resented by her husband and his relatives and therefore diminish her chances of happiness. Her parenthetical comment, suggesting that there will be plenty of other reasons for unhappiness, indicates that her hopes for the marriage were not high. Several years later, Delaval summed up the blessings of marriage as 'the kindnesse of a husband and the unspeakeable comfort of haveing pay'd my creditors' (p. 209), suggesting by her choice of vocabulary that the greater comfort was financial!

Delaval did indeed win her thousand pounds as money of which she could be the 'mistress', after a 'long and firce argument' with her aunt:

> At length I told my aunt that I was very sure if my grandmother knew what pass'd on earth, I was very sure she wou'd be much displeased with her for intending to hinder me from being misstress of what my deare grandmother had given me upon her death bed. My aunt (who was extreamly good natured), being moved by those word's, shed some tear's and Imediately gave order that the thousand pound shou'd be pay'd me. (p. 69)

By evoking the emotional force of her late grandmother, Delaval successfully (though perhaps slightly unscrupulously) wins the element of financial selfhood bequeathed to her. What is particularly fascinating is that all this wheeling and dealing is described in the context of apparently spiritual meditations. In attempting to explain herself to God, Delaval finds that earthly considerations such as money and marriage play a prominent role. The discourse of her spiritual 'life' and her financial 'living' significantly overlap. Her references to 'fortune' interchangeably signify providence or fate on the one hand, and her finances on the other; the use of the term 'misstress' in her meditations implies a woman in charge of her own conscience as well as her own purse. When she refers to herself in the final prayer of her text as a 'poor bankrupt sinner' (p. 212), one suspects that the mercantile language is both metaphorical and real. Her ideal life is summed up a few pages earlier as to 'be able to pay all my debts and satisfy the murmering whispers of my conscience' (p. 186). Since both God and the thousand pounds have been mentioned within the same paragraph, the boundary between devotional and financial concerns remains unclear.

Life and living, then, in the sense of personal selfhood and material existence, are surprisingly difficult to disentangle in the lives – both lived and written – of these seventeenth-century women. The closeness of identity and money was a particular pressure for women since, as we have seen, a wife herself represented a purchase or investment as well as an individual with needs and talents. As Lady Anne Clifford found in the early years of the seventeenth century, a woman's money and her body were very closely identified. Clifford recorded in her *Diary* the struggles she had to maintain her rightful claim to the inheritance of her father's lands in Westmorland, not only against her uncles to whom they had been willed, but also against her husband who wanted her to give up the claim in return for a quick cash settlement. Clifford's defiance of her husband's wishes thus represented not only a continuing assertion of her own family identity after marriage, but also the denial of immediate funds to pay her husband's debts. Not surprisingly, the legal and financial arguments concerning Clifford's lands – which took her several times to the court of the King in London – are frequently mentioned in her diary. A sequence of entries in April 1617 suggest the regularity of their discussions:

> The 16th my lord and I had much talk about these businesses, he urging me still to go to London to sign and seal; but I told him that my promise so far passed to my brother and all the world that I would never do it, whatever became of me and mine. Upon the 17th, in the morning, my lord told me he was resolved never to move me more in these businesses, because he saw how fully I was bent.[6]

There are two interesting points to notice here. First, the euphemism for the whole financial discussion and its consequences is 'these businesses' or, elsewhere in the diary, 'the business' or simply 'matters', words which refer very properly to financial issues, but in their repeated use take on something of a disdainful tone, as though hiding a multitude of sins and certainly suggesting physical as well as monetary mysteries. Second, Clifford's refusal to sign away her claim is seen as crucial to her individuality, to the future of 'me and mine'. Her life, as well as her living, is potentially at stake here in this 'business'.

As the days of April 1617 proceed in Anne Clifford's diary account, the debate with her husband over the 'business' becomes tangled up with spiritual matters and its impact becomes more intimate:

> The 19th I signed 33 letters with my own hand which I sent by him to the tenants in Westmorland. The same night, my lord and I had much talk of and persuaded me to these businesses, but I would not, and yet I told him I was in perfect charity with all the world. All this Lent I ate flesh and observed no day but Good Friday. The 20th, being Easter Day, my lord and I and Tom Glenham, and most of the

folk, received the communion by Mr. Ran. Yet in the afternoon my lord and I had a great falling out, Matthew continuing still to do me all the ill office he could with my lord. All this time I wore my white satin gown and my white waistcoat.

The 22nd he came to dine abroad in the great chamber; this night we played burley break upon the bowling green. The 23rd Lord Clanricarde came hither. After they were gone, my lord and I and Tom Glenham went to Mr. Lune's house to see the fine flowers that is in the garden. This night my lord should have lain with me, but he and I fell out about matters. (p. 48)

In this last entry we observe the effect that the financial disagreements have upon the physical relationship of husband and wife. Anne Clifford refrains from giving her consent to her husband's financial proposal by which her individual inheritance would be denied; he, in turn, refrains from sexual intercourse with her. Her money and her body are part of the same equation; the preservation of one implies the denial of the other. This is an extreme example of the significance of Clifford's physical identity throughout the quarrels; it is clearly no coincidence, for example, that she wears her best white clothes despite the 'great falling out' on Easter Day. The whiteness suggests purity and innocence, as does Clifford's comment that although she refused to accept her husband's financial plan, she 'was in perfect charity with all the world'. Through the plainness of Clifford's prose shines an image of the wronged heroine.

It is likely that Clifford remained a heroine mainly to herself, since her diary lay unpublished until this century. Like Martha Moulsworth and Elizabeth Delaval, Anne Clifford wrote primarily for her own satisfaction or that of her conscience. Their lives took shape in their own words; their identities were formed in the interplay of different discourses – of money, marriage, devotion and the body. But such seventeenth-century expressions of female life and living were not always private. One notorious woman, Mary Carleton or the 'German Princess', wrote and published an account of her life in order to set her own record straight, since she was the subject of numerous rumours and pamphlets during her trial for bigamy in the 1660s. It is difficult to establish the truth of Carleton's adventures, for the interwoven narratives of her life are complex, but in *The Case of Madam Mary Carleton* (1663) she defends herself against the charge of tricking John Carleton into marrying her under false pretences. It is clear that she began to make a fiction of her life fairly early on, sending false letters in order to trap those men who were themselves pretending to be rich or noble in order to win her hand in marriage. As she comments, it is a 'received principle of justice' that 'to deceive the deceiver, is no deceit'.[7] She had arrived in England as the daughter of the German lawyer, Henry van Wolway, a surname which, when used by her, 'suffered by another lewder imposture and allusory sound of "de vulva"' (p. 136). While Carleton built up false ideas of her riches, those around her began to create stories and images of her based upon a *double entendre* perceived in her name. The physicality of the jibe is interesting; once again, the rich woman who, however temporarily, has power over men through her (supposed) wealth is identified suggestively with her female body – in this case, specifically the genitalia. When another woman observes to Carleton that it is clear why Mr Carleton loves her – for her 'great parts and endowments' (p. 137) – do these terms refer to her body, or her fortune, or both?

In her lively history of the courtship of Mr Carleton, written to defend herself but also to earn some money by the publication, Mary Carleton plays teasingly with the closeness of fact to fiction, borrowing in particular from the style and traditions of romance. The following passage is notable for its intrigues and romantic clichés, played off against reality by a knowing narrator:

> After many visits passed betwixt Mr Carleton and myself, old Mr Carleton and Mr King came to me and very earnestly pressed the dispatch of the marriage, and that I would be pleased to give my assent, setting forth with all the qualities and great sufficiencies of that noble person, as they pleased to style him. I knew what made them so urgent, for they had now seen the answers I had received by the post, by which I was certified of the receipt of mine, and that accordingly some thousands of crowns should be remitted instantly to London, and coach and horses sent by the next shipping with other things I had sent for. And to reinforce this their commenda-mus the more effectually, they acquainted me that if I did not presently grant the suit and their request, Mr Carleton was so far in love with me that he would make away with himself, or presently travel beyond the sea and see England no more.
>
> I cannot deny but that I could hardly forbear smiling to see how serious these elders and brokers were in this 'love-killing' story. But keeping to my business, after some demurs and demands, I seemed not to consent, and then they began passionately urging me with other stories . . . (pp. 136–7)

The narrator's enjoyment in this extract derives from the ironic fact that she is the hard-headed 'business' woman here, while the men are indulging in fictions of passion such as threatened suicide and other 'stories'. Those who were supposed to be 'brokers' in the marriage market were turned by her manipulation into brokers of fiction rather than wealth. As the passage ends, they are forced into what Carleton refers to as 'stories', which in the subsequent paragraph turn out to be pretended grandeur, lordly disguises for Mr Carleton, and false fables of his wealthy connections which are neatly epitomised in a 'rich box of sweetmeats' (p. 137) presented to Mary Carleton. Like too many sweet-meats, the liaison with Mr Carleton is only superficially rich and sweet, and very soon leads to sour sickness for both parties.

What was Mary Carleton's life? It is remembered as a glorious fiction, criminal in its own deceptions but fascinating in its revelation of the manipulation, dishonesty and self-deception of traditional courtship. She played the market and, in winning her court case, brought down some of the other 'brokers'. But she lost a great deal, too, as *The Case of Madam Mary Carleton* demonstrates. When Mr Carleton senior decided to prosecute his daughter-in-law, she was 'by a warrant dragged forth' from her lodgings and

> by him and his agents divested and stripped of all my clothes, and plundered of all my jewels and my money, my very bodice and a pair of silk stockings being also pulled from me; and in a strange array carried before a justice. (p. 141)

Despite her former control of the situation, Carleton is shown to be vulnerable in a system which was run by men and showed little regard for women's integrity. The 'German princess' has become a whore in the eyes of the law; she neither needs, nor deserves, the dignity of clothes and is 'stripped' physically and financially. This incident in her account prefigures the end of Carleton's life. Unable to keep her own money or her earnings as a writer, since though rejected by her husband she was still legally married and therefore all her belongings were his, she was ultimately arrested for theft and hanged at Tyburn in 1673.

The dramatic case of Mary Carleton makes it clear that autobiographical writing can create a life-fiction even while the historical persona has little living and in the end loses her life. Text and reality, fact and fable are blurred here in a most complex way; in the end, though, Carleton had her own life taken from her. The most urgent of all autobiographical texts are those which were written to save a life, whether from slander or accusation or death itself. One seventeenth-century woman who wrote her autobiography in self-defence, but whose case in most other ways bears little resemblance to Mary Carleton's, was the Fifth Monarchist prophet, Anna Trapnel. Where Carleton used the sensational style of the romance to glorify her own experience, Trapnel used a biblical manner in writing her *Report and Plea* (1654) to honour God by demonstrating his actions at work in her. During the account of the day of her trial for illegal prophesying and preaching, the parallels between her experience and Christ's on Good Friday frame the narrative:

> So I went, the officer coming for me; and as I went along the street, I had followed me abundance of all manner of people, men and women, boys and girls, which crowded after me. And some pulled me by the arms, and stared me in the face, making wry faces at me, and saying, 'How do you now? How is it with you now?'. And thus they mocked and derided at me as I went to the sessions. But I was never in such a blessed self-denying lamb-like frame of spirit in my life as then; I had such lovely apprehensions of Christ's sufferings, and of that scripture which saith, 'He went as a sheep, dumb before the shearers, he opened not his mouth; and when reviled, he reviled not again'. The Lord kept me also, so that I went silent to the sessions-house, which was much thronged with people: some said the sessions-house was never so filled since it was a sessions-house. So that I was a gazing-stock for all sorts of people, but I praise the Lord, this did not daunt me, nor a great deal more that I suffered that day, for the eternal grace of Jehovah surrounded me, and kept me from harm. So way was made for me to draw near to the table, which stood lower than the justices; and round the table sat the lawyers and others that attended them, and I with my friends that went with me stood by the lawyers, and the justices leaned over a rail, which railed them in together . . . I had the [ ] courage to look my accusers in the face, which was no carnal boldness, though they called it so.[8]

The subjective perspective of autobiography – one of its essential strengths – is vividly shown here in Trapnel's account of the day of her trial. With her we see at close range the 'wry faces' of the onlookers who jeered from the sidelines as she walked along the street, and hear both the external taunts of the crowd and the internal biblical promptings of her spirit. While others take her for a madwoman, a witch or at the very least a female showing 'carnal boldness', she perceives her mood as serenely holy and her attitude as a demonstration of 'courage'. The norms of perspective are reversed. She may be a 'gazing-stock' for those who throng the sessions-house, but she knows that her steady gaze into the faces of those who accuse her is the more significant and disruptive angle of vision. Although she is the prisoner, her description of the courtroom instead places the justices in a sort of prison, 'railed in together', while she stands in the company of her friends. Again we are tempted to ask, what is a woman's life? To what extent does Trapnel's account challenge not only the stereotypes of sane and insane – particularly when those categories are used to characterise women – but also the relative importance of the physical and the spiritual? There is nothing 'carnal' about this experience, except the misinterpretation of it; life for her rests on an eternal rather than a mortal ground, and the idea of making a 'living' seems supremely irrelevant.

Seventeenth-century women can have this uncomfortable effect of jarring our assumptions. This is not only because of their sometimes startlingly spiritual perspective, but also on account of the boldness of their actions and the firm clarity of their justification of them in writing. Take, for example, the case of Anne Wentworth, who left her husband in the late 1670s in order to follow her vocation to write about her experience of God. In her autobiographical text, *The Vindication of Anne Wentworth*, she unhesitatingly justifies her action – seen by others as disobedient to the norms of femininity in marriage, religion and writing – and she does so by referring to inner values and a calling to higher tasks than earthly categories can comprehend:

> And however I am censured and reproached by persons who judge only according to *outward appearance*, but not *righteous judgment*, that I have unduly left my husband, I do for the satisfaction of all plain-hearted ones that may be offended at their reports herein declare, *first*: that it would be very easy for me, from the great law of *self-preservation* to justify my present absence from my earthly husband ...Forasmuch as the natural constitution of my mind and body, being both considered, *he* has in his barbarous actions towards me a many times over done such things as not only in the *spirit* of them will be one day judged a murdering of, but had long since *really* proved so, if God had not wonderfully supported and preserved me. But...I will not urge anything of this nature as my defence upon this occasion, having learnt through the mercy of God not to be afraid of him who can only kill the *body*, but can do no more. I do therefore *secondly*, in the fear of him who can kill both *soul* and *body*,

further declare that I was forced to fly to preserve a life more precious than this natural one, and that it was necessary to the peace of my soul to absent myself from my earthly husband in obedience to my heavenly bridegroom, who called and commanded me, in a way too terrible, too powerful to be denied, to undertake and finish a work which my earthly husband in a most cruel manner hindered me from performing, seizing and running away with my writings.[9]

The rationality of Wentworth's argument is plainly to be seen: the first defence of her separation from her husband is the preservation of her body in the face of his 'barbarous actions', while her second reason for absenting herself is the 'peace' and safety of her soul, thus preserving 'a life more precious than this natural one'. As is so often the case in autobiographical writings, there are competing selves in this account – her earthly and her spiritual lives – which are matched by an 'earthly husband' and a 'heavenly bridegroom'. Wentworth has no doubt about where her priorities lie, especially when what might be seen as her ultimate identity – her writing – is allied with the higher or spiritual self. Although writing is itself a physical action, its significance for Wentworth is its ability to give expression to her 'heavenly vision' (p. 187). The last straw in her marriage was her husband's 'seizing and running away with my writings'; this, and not his physical violence, is described as a 'most cruel' act. It is evident from Wentworth's rhetoric that for her the fundamental self-definition came not from marriage, nor from other material concerns such as money, but from her capacity to write, not only about things divine but also, as in the case of her *Vindication*, about herself in relation to God.

It is possible to suggest, therefore, that self-writing *was* living for some seventeenth-century Englishwomen; Anne Wentworth found identity in the pages of her own rescued works, just as Martha Moulsworth surveyed her life and came to a sense of purpose in widowhood through her 'Memorandum'. There are lives, and there are livings, but perhaps these are only brought together in the written and constructed self of an autobiographical text. One of the finest instances of this in the seventeenth century is the memoir of Margaret Cavendish, written for no other apparent purpose than the sheer pleasure of self-inscription. There is no religious motive, no need to earn a living by her account (though she did publish it) and no obvious motive of self-defence in the face of false accusation. Cavendish's *True Relation of my Birth, Breeding and Life* (1656) is the first known published secular autobiography by a woman, and there is in it a genuine sense of creating a life through the words of her historical sketches and character analyses; it is an account driven by a preoccupation with how 'after ages' will regard her, and intent upon influencing that process. The most striking passage of Cavendish's autobiography is, appropriately for my last example, its concluding lines:

> But I hope my readers will not think me vain for writing my life, since there have been many that have done the like, as Caesar, Ovid and many more, both men and women, and I know no reason I may not do it as well as they. But I verily believe some censuring readers will scornfully say, why hath this lady writ her own life? Since none cares to know whose daughter she was, or whose wife she is, or how she was bred, or what fortunes she had, or how she lived, or what humour or disposition she was of? I answer that it is true, that 'tis no purpose to the readers but it is to the authoress, because I write it for my own sake, not theirs. Neither did I intend this piece for to delight, but to divulge; not to please the fancy, but to tell the truth. Lest after-ages should mistake, in not knowing I was daughter to one Master Lucas of St John's near Colchester in Essex, second wife to the Lord Marquis of Newcastle; for my lord having had two wives, I might easily have been mistaken, especially if I should die and my lord marry again.[10]

This passage betrays the uncertainty inherent in the enterprise of auto-biography, particularly for a woman. Is she being hideously vain in writing her own life story? Should anyone else really wish to read the detail of her life? Does she seek fame, or any kind of posthumous iden-tity, or is it just for personal satisfaction? If so, then why publish it? These are the circles of self-doubt which apparently sapped the confidence of even this most determined of seventeenth-century autobiographers. And the concerns expressed here echo those of all the women whose work I have been considering. Like Martha Moulsworth, Cavendish, though obviously of an independent mind, has to fall back on her father and husband to offer the recognisable frame for her identity: 'whose daughter she was, or whose wife she is ... daughter to one Master Lucas of St John's ... second wife to the Lord Marquis of Newcastle'. Like Elizabeth Delaval, Cavendish refers to her 'fortunes' with an ambiguity which reminds us of the close interrelation of money and providence for women in the seventeenth century. Just as Anne Clifford had held on to her claim to the lands of her own family, so Cavendish here insists upon her own identity; she does not wish to disappear into oblivion as an unknown and unnamed wife of the Marquis of New-castle. Like Mary Carleton and Anna Trapnel, who in their contrasting ways defended their own perspective on the events of their own lives, Cavendish, too, asserts that she writes 'to tell the truth' about 'her own life'. And yet Cavendish maintains to the end that she writes 'for my own sake', suggesting that, like Anne Wentworth, she is most fully herself when in the act of writing.

What, then, was Cavendish's 'own life'? It was a peculiar mixture of pride and modesty, likened to the lives of Caesar and Ovid but in the next breath denied any significance to any reader beyond the author herself. Clearly her own material 'living' formed an important part of her self-definition; Cavendish began life as the daughter of a mere 'Master' but ended up as the wife of a Marquis, more than a step up in the world. The material and the physical – her 'humour' and 'disposition' – mingle in the making of this life; but without its written definition, she might, the last lines imply, dwindle to dust without any distinctive identity. As it was, the text of her life might have bitten the dust had no copies survived of the first edition of *Nature's Pictures*, the collection in which her memoir appeared; for in the second edition, the autobiography is no longer included. A woman's life, like her living, can be a fleeting phenomenon.

These autobiographical texts, though enormously varied across the spectrum from semi-fictional romance to spiritual defence or meditation, have suggested some fundamental common ideas on women, autobiographical writing and the material world. We have witnessed women in the marketplace as marriageable virgins and as writers, but also in the home, the church and the courtroom. We have seen their dependence on money but also their determination to gain some power over material resources; on the other hand, we have observed how for many women the release into spirituality or textuality represented an alternative or fuller sense of living. In the interplay of gender, materiality and the textual 'life', early modern women betray themselves into immortality.

Acknowledgement

I am grateful to colleagues in Groningen, and at Sheffield Hallam University, for constructive comments on this essay.

Notes

1 For an account of women's writing in early modern Britain, in the context of religion, education, literacy and literary production, see *Women and Literature in Britain 1500–1700*, ed. Helen Wilcox (Cambridge, 1996).

2 Among the recent studies of women's autobiographical writing in this period, see Elspeth Graham, 'Women's Writing and the Self', in *Women and Literature in Britain 1500–1700* pp. 209–33, and Sheila Ottway, *Desiring Disencumbrance: The Representation of the Self in Autobiographical Writings by Seventeenth-Century Englishwomen* (University of Groningen PhD thesis, 1998).

3 Margaret Cavendish, Duchess of Newcastle, *A True Relation of my Birth, Breeding and Life* (1656) in *Her Own Life: Autobiographical Writings by Seventeenth-century Englishwomen*, ed. Elspeth Graham, Hilary Hinds, Elaine Hobby and Helen Wilcox (London, 1989) p. 98.

4 'The Memorandum of Martha Moulsworth, Widow' (1632), modernised text from *'The Birthday of Myself': Martha Moulsworth, Renaissance Poet*, ed. Ann Depas-Orange and Robert C. Evans (Princeton, 1996) pp. 11–14.

5 *The Meditations of Lady Elizabeth Delaval*, ed. Douglas G. Greene, Surtees Society vol. CXC (Gateshead, 1978) pp. 68–9.

6 Lady Anne Clifford, *Diary*, in *Her Own Life*, p. 48. The full text of her diaries is available in *The Diaries of Lady Anne Clifford*, ed. D.J.H. Clifford (London, 1990).

7 *The Case of Madam Mary Carleton* (1663), in *Her Own Life*, p. 135.

8 *Anna Trapnel's Report and Plea* (1654), in *Her Own Life*, pp. 78–9.

9 *A Vindication of Anne Wentworth* (1677), in *Her Own Life*, pp. 186–7.

10 *Her Own Life*, pp. 98–9.

8
The Two Pilgrimages of the Laureate of Ashover, Leonard Wheatcroft

Cedric C. Brown

The broad development through the fifteenth to eighteenth centuries of diaries and autobiographies, that is to say, forms of personal record of shorter and longer perspective, is well known. It is understood, for example, that the personal record developed slowly out of annals of memorable events, that the practice moved from clerics to lay persons, especially in towns, that spiritual self-recording became common during and after the religious struggles of the sixteenth century, and that there is a thin line between various kinds of accounting and self-accounting, since both can be about the use of the talents or 'the infinite benefit of daily Examination: Comparing to a Merchant, keeping his books, to see whither he thrived...'[1] Many kinds of record attempt to clarify what happened in the past in order better to gear up for the future. But, as the present volume witnesses, there are many kinds of personal record, many more than can be comfortably assimilated to one model, and in any period more dispersed or differently figured kinds of self-recording can be found. This essay is about a remarkable and little known set of seventeenth-century documents which may serve, at the least, to make us think about the relationship of 'private' self-representations and 'public' self-presentations, and about the effects of the wide distribution through print of collections of socially useful or self-improving kinds of writing, books of 'wit' in verse and exemplary letters.

My writer is a Derbyshire yeoman who was born in 1627 and died in 1706, Leonard Wheatcroft. Two personal narratives of his survive in manuscript, an autobiography and a story of courtship.[2] I want to use the courtship narrative as much as the autobiography, because it is the livelier of the two, and presents more vividly the problem which I wish to identify in the present collection on forms of autobiography. That is to say, although the courtship narrative concerns private record, it is manifestly also public entertainment, in the sense that Wheatcroft's project in the manuscript miscellany book which contains it is to display writing and his writings, to be a performer and a laureate in his country community. There is little to connect these texts with those models of spiritual pilgrimage we tend to associate with the period; although, as we shall see, there are elements of accounting in them, these autobiographical narratives have modes which are unmistakably social, offering communal entertainment and instruction.

The running head at the top of the pages of his two miscellany books hints that he set out to peddle his wares, or at least to imitate those who do: 'Cum you galants looke and by / I am myrth and melody'. This looks like broadside selling. There is a lot of mirth and melody in his collection, and Wheatcroft himself often refers to good alehouse company. Written in this spirit, one of the first poems forms a heading to his book:

> to the vewers & standers by.
>
> Cum you galants looke and by:
> Heare is mirth and melody:
> Heare is epegrimes to learne
> And perrose you will not Scorne:
> Heare is lines on euery sort
> That will make your sweethart sport:
> If you please to cum and by:
> My name is mirth and melody:
> Sum my sweethart did mee send:
> Others to her I did lend:
> Which will make you for to smille:
> To heare how Lovers did begille:
> Other things (you may heare find)
> [ ] for every state and kind:
> . . .
> And youre eyes for to behould:
> And let out teares, a thousand fould:
> Yea thy sences euery on:
> He will rashly beat up'on:
> To shoote he has, such a brave art:

> That If he draw he hites the hart:
> Hee'le make you sigh allmost to die
> . Conclusion mirth and melody.
> ('Gallants', third page, first pagination)[3]

The text presents Wheatcroft as the instructor-wit, especially to gallant youth, of his part of Derbyshire, a communicator of the social benefits of knowing discourse to his country, and also as the bard to speak to the feelings of the neighbouring society.

Thanks to a comic story Wheatcroft himself tells in his autobiography, we know that his capacity for laureate self-display and the grand gesture was recognised by the local community. In 1691 he wrote a semi-allegorical poem celebrating the Glorious Revolution sited at Ashover's 'fabrick', a gritstone quarry on the hill above his Parnassus (p. 96). This pretentious text was mocked by a rival wit of another village, Tupton, so, when Wheatcroft was in his late sixties, a group of gentlemen arranged a contest between the two rustic bards:

> there did I challeng him to whake with me to Parnishus Hill, but we both mising our way, we chanced to light of an all-hous, and after we had drunk awhile we fell into discors concerning the 9 Muses which he could not name, naithe could he tell from whence the[y] came, or what the[y] had done, or what the[y] might doe, so I in the audienc of all the companey gave them their right names and all their right titles, where upon the[y] decked my head round with lorill branches to the great vexation of my antagonist, Ouldham. So ever since I am called the Black Poet.[4]

Bacchus is father to poetry. It can also be seen that Wheatcroft was a colourful man, one who presented himself in style. He was concerned to show authority, in this case as possessing a direct line to the ancient founts of poetry, this in turn a badge of the kind of education which gave social advantages to the upper echelons of society. The authority as educator conveyed by this tale fits the fact that he was for many years parish clerk and register at Ashover, and briefly village schoolmaster.

His background was artisan. His father was a tailor, and that was his trade which he taught to four of his younger brothers after his father died. But he was also versatile: he learned to plant hedges and orchards and was employed by local gentry for that purpose; he could build, was a joiner, and make furniture; he tells us in passing that he knew how to tune the virginals, and for several years he was involved as stakeholder in lead mining at Youlgreave. His son tells us that he had a good singing voice, leading in church, and he was a keen bellringer.[5] And, apart from the manuscript books I have referred to, of poems and 'perrose', he apparently wrote a catechism entitled 'A Free Gift to Ashover Free School' in 1673 and a book called 'The Bright Star of Love appearing for Bachelors', twenty-one years in the making, he said, and (wishful thinking) ready for the press, as well as various other things I shall mention.[6] For a yeoman, there is no social stigma in thinking of print; indeed, as we shall see, much of the early impulse for writing came from popular printed miscellanies. He also refers grandly to his library, an obvious mark of social status and a significant expense.

Yet his fortunes collapsed: somehow, perhaps through over-reaching and his natural expansiveness, he lost the house and small lands he had inherited, spent years in penury, some of them in exile at Bolsover, was three times imprisoned for debt, and only gradually recovered his fortunes and social status in later life, by which time his wife had helped the domestic economy by bringing up the nine surviving children and becoming an ale-house keeper: their house was called 'The Hand and Shears', referring to Leonard's tailoring.

But let us look at the two texts themselves, and first at the autobiography. The little calf-bound volume is called 'A History of the Life and Pilgrimage of Leonard Wheatcroft of Ashover'. Pilgrimage might in other circumstances imply a spiritual journey. There are, as I have said, some elements of self-assessment, but this factual narrative of notable events and activities in his life lacks agonies of conscience. The narrative is retrospective, written in the past tense, and there is something of a shape in it to be traced through financial distress in middle life to slow recovery from the very end of the 1660s and the 1670s, during which time the family could once more take up their position in Ashover village society. In the end, Wheatcroft could record the fact that he had contributed to the fund for rebuilding St Pauls, a symbol of having arrived. The last parts of the narrative also tell with satisfaction how he placed his children in service and apprenticeships.

There is nothing about his childhood or early education, and detail tends to get fuller from about 1670. Wheatcroft's courtship narrative uses many military metaphors, but here in his History he says nothing about his soldiery in the local militia in the 1650s, for the simple reason that he had described that in yet another book, now lost.[7] What strikes one throughout this book is the emphasis on positive things to remember and the frequency of mention of some particular kinds of activity.

He likes to record plantings of hedges and trees and orchards and working in the gardens of the gentry. This activity features as much as tailoring, and it often took him away from home. It was probably a good earner and it put him in contact with higher social classes.

He always mentions, too, the writing of epitaphs on the occasion of local deaths, sometimes of people of rank. There are also other presentation verses to local nobility, for a birthday tribute, for example, for which he was sometimes paid.[8] This goes together with his references to his other writings, and it was clearly crucial to his role to be the occasional poet to his local community. He was, as it were, the Ben Jonson of his county.

He always records going to wakes and fairs, many of them local, some quite a way away. Community and family merry-making are immensely important, expressing a strong cultural imperative. (It may be also that we should associate fairs with the peddling of literary wares, in the form of cheap broadsides.) Wheatcroft had been a soldier for parliament, but clearly this is no strict Puritan – there is a strongly embedded local culture not greatly moved, it would seem, by political change.

Then, as I have already said, he has an obsession about journeys. He is always going off somewhere, even to London (perhaps four times in his life), perhaps to plant, perhaps to see distant kinsfolk, perhaps to social events, or to assizes, or an election in Chesterfield, sometimes simply to find someone who would lend him money. In some ways, the association of travelling with personal record is typical of the period: journeys, like wars, often provided a stimulus for writing down one's experiences. They were also expensive, so were occasions for reckonings. But Wheatcroft has an interestingly complicated attitude towards all his travelling. At moments he recognises that it is extravagant. Concerning his life about 1670, a time of general stocktaking, he wrote:

> but as for my daly travills and idle jorneys to and fro the[y] came to above 550 miles a yeare, which I believe I went above 20 years togather, as my accounts shewed me. (p. 87)

He often totted up mileages; however costly, journeys also denoted achievements. Not everyone in the village community, presumably, travelled quite so often as he. There may also be something of the business of bringing back to his local community the news of the places he had visited.[9] We will see the preoccupation with travelling as expense and achievement in both History and courtship narrative.

Here are two characteristic extracts from the autobiography:

Jan. 1 I fell to worke and rought very hard till Feb. 19 that day I fetched Mr Dakin a vine and a cheritree from Batmans so I continued gardening till Feb 28. that day being Chesterfeild faire day I like a bad lad, went to it, and stayd there all night. Mar. 7 being fassen-Tusday my wife and I according to ould custum went over to Winster to se our relations and freinds.

Mar. 11 I had sum notions of being clarke of the Church of Workesworth and the 16 day I went over to speak to sum freinds, of whon I had good hopes of the place. But preveled not. June 4 1671 my wife and I went over unto Winster againe to the christning of my Brother Roland's child whose name was Eliz after that I rought hard of my Trade till July 15 that day I thought to have bult me a house at Sir William's well, but our goodly parson Obediah – that small profit – would not suffer it, because I had puld downe his fathers intacke in Asher hill. Then did I fall on in my owne ground, and begun to rid for a house stid in a place which I call now by the name of Hockley. But how that will prove, you shall heare more hear-after.

So going on with my trade one whit and gardenin, and ridding another I rit downe these questions following.

1. What is the reason that one man is not like another in phisogmony.
2. What metell is the sight of the eye made of.
3. Why have men beardes and women none? . . . etc. (pp. 87–8)

One can see in this the concern to register two longish spells at his trade, nine days gardening out of the village, a binge at Chesterfield, family festivities and his desire to secure a clerkship again. Also some of his better humour – the vicar who had denied him is dismissed as a minor prophet – and some of his incorrigible digressive showing off – the thirty conundrums of which I have mercifully reproduced only the first three, have no part in the narrative, but Wheatcroft spills on to paper all the fruits of his (or someone's) wit. He is a purveyor of entertainment to his community.

The second extract comes a little later:

Till the 73 and all that yeare I did many things of noate, I writ a Catechisme, titeled a free gift to Ashover free Schoole, and other memoruble things, now in writing. Mar. 26 seventy three, I set that sick-amore at the Rodd and upon May 2 after I came to my ould house againe, which I had morgaged to one Antoney Tourner, But like a good lad I sould it about 2 years after, for I was in sum debt and nothing would serve crediters but money money.

And more-over the same day I came to my house againe my wife fell a labour and was brought abed of a son whose name we called Sollomon. May 2 73 so then being arived at my owne house againe, I seased to buld me a new one, but repared my ould one which was left foule out of order so as five pounds did not make it good againe, for one beay of my barne was fallen besides windowes swat all in peeces.

In 74 I began to write another booke called The Memorys Recreation, but if I live to make it out, you will all be better satisfied and I cumended however the coppy of it you may find among the rest of my writings in my libry.

Octo. 2 being Friday 1675 was my wife brought abed of a doughter whose name we call Maddam Sarah. My wife continued all this while abrewing and I rought and did what I could so that it pleased god we did a little recover againe and got ourselves and children close, paid sum debt, and so by degrees gained our freinds again. (p. 90)

Here we have the pride in writing, the concern to have or build one's own house, births of children (including a wry name for a daughter, Madam Sarah), and above all the importance of gaining an unembarrassed social position. He also treats the readers of his History as a general admiring audience of all his literary works – 'you will all be better satisfied and I cumended however the coppy of it you may find among the rest of my writings in my libry.' Even the factual History turns from time to time into a laureate work.

The courtship narrative is not separately bound, but comes in the mixed materials of the second book of Wheatcroft's manuscript miscellany, 'Cum you galants'. That context is important in assessing the way it is written. The implication is that this personal narrative about his two-year courtship of Elizabeth Hawley and marriage to her is being presented as amusing and exemplary, and indeed we shall find it full of romantic and heroic inflections. So there is some remaking of life materials into marketable copy, and the already subdued note of self-examination in the autobiography is here completely absent: this is semi-fictionalised celebration. The mood can be caught from the opening paragraph:

Well, gentle reader, you see how I have extolled not only one part but every part of woman, and why I did so was because I loved every part; but finding none as yet to seal an impression upon, I could not withhold my muses till they had uttered what Love was; and finding it to be both a friend, a fire, a heaven, a hell, how could I do any less than express all I know of it? Nay, could I express how it hath betrayed me, ye would even wonder that I could have patience to write it down. Yet for your further recreation I will declare what fortune I had in my pilgrimage towards this town called Woman.

First, I having mustered and called up my forces many a time, I never could advance towards her for lack of something or other ...
(p. 36)

So he keys his narrative back into early parts of the miscellany, in which amatory verses predominate (thereby implying that he had been a bit of a gallant himself), protests the importunity of the muses, and narrates his personal capture of the Town called Woman. In this the word 'pilgrimage' simply means an arduous quest, as it probably does in the History. This is also a retrospective telling, dated soon after the events.

The courtship narrative has a most unexpected beginning, which gives details of one of his journeys to London to visit a brother and see the sights there. This is presented as a false start, the idea apparently being that he first set about advancing towards the Town called Woman the wrong way by seeking his romantic fortune in the capital, only to find, when he got back, that the destined damsel was waiting nearer home in Winster. Winster is a village some eight miles away on the other side of Darley Dale. Although this device is comic-chivalric and loyal to his country, it is also to do with his tendency to want to tell his countrymen the big sights he has seen elsewhere. So he relates how he visited all the ancient tombs of kings and lords in Westminster Abbey: 'But before I came away I made bold to sit me down in the king's chair, etc.,' saw 'the Lord Protector and his royal train', went to the Tower, and of course viewed the fashions of the court, 'and withal the beauteous ladies in their balconies, and I not espying one to please my fancy did resolve to see my own country again ...' (pp. 39–40).

The details of courtship and marriage form an extremely valuable record of social history. There is a good picture of the length of negotiations between Leonard, and his friends, and Elizabeth's father and his advisers about the marriage settlement, a negotiation which came apart several times and was only finally fixed in a meeting of a day and a half perilously near the time of the wedding. Even more, there is a description of the festivities of the wedding day itself and the week or so after, a truly wonderful description forming the triumphant climax to the narration. It is an account probably unequalled for this provincial society in any other document.

As a structure, the narrative has a shape easy to describe: Wheatcroft engages in a protracted negotiation over two years, so there are difficulties to be overcome, all of which makes the final triumph seem the greater and gives ample opportunity for display of know-how in the business of conducting a difficult courtship campaign. This also means that he can draw out of the earlier miscellany various exemplary and personal letters and poems, which are now shown in action.

As a first example we might take his opening letter:

But finding none of my tongue-shot would do any execution, I charged my fiery heart-piece with these ensuing lines, saying

Sweet Betty,
If I could have as good passage as these my lines, I would be better messenger of my thoughts than my lines can here express; but, dear Love, the secrets of my love is sealed up here in my heart, which none can see but those starlike eyes of yours, nor shall any know but your kindness. O let me not then languish in the lingering hopes of my desires, but hasten my comforts in the only answer of your free consent, for trial cannot let you doubt of my constant love and loyalty, and my love shall be sworn for the security of my truth. O then give truth the reward of trial, and love the regard of truth, and do not now defer the sentence of justice. O then let me live or die in your judgement, for now am I imprisoned in your splendent beauty ... and bound in the bonds of your service, and live in the hopes of your favour, in which I shall rest ever and only happy with you,

Yours though not yours,
Leonard Wheatcroft
(pp. 42–3)

Romantic expression was common in real courtship, as we know from many sources, and there is no clear line between fictional and real-life constructions, but one cannot help wondering in this case, if such a letter were really sent, how Elizabeth had received it. Was she simply susceptible to the powers of Leonard's imported rhetoric?

There are other expressions of similar sort, like the one of 22 August 1656, beginning:

> O my dearest Love,
> How long must I wait at the pool of your Bethesda (all besmeared with sorrow) ere I hear that sweet echo from you, I am yours? ...
> (p. 45)

This and other letters bear dates, so presumably something must indeed have been sent. Wheatcroft also includes some replies from Elizabeth, which are predictably much plainer in style. Also 'real' rather than merely fictional are various personalised verses, like those which are built on her name. There is, for example, a laboured triple acrostic on Elizabeth Hawley's name. Because of its extreme difficulty, this is a very proof of wit, though an appalling poem. But sometimes the poems he sent were borrowed or adapted from elsewhere, and sound altogether smoother, as in this next example:

> ... I found so small an alteration in them [her communications] that I could noways withhold my tongue and pen from warbling forth these ensuing lines

> A poem on my Love to her praise

> I have a mistress for perfections rare:
> In every eye, but in my thoughts most fair.
> Like tapers on the altar shines her eyes,
> Her breath is the perfume of sacrifice,
> And whereso'er my fancy would begin
> Still her perfection lets religion in.
> There grows upon her cheek a pure carnation,
> Come taste a kiss, O sweet O sweet temptation.
> (pp. 51–2)

Such poems came from other circles, this one from Thomas Randolph, in fact. It appeared in at least one mid-seventeenth-century miscellany.[10] So this is a second- or third-hand warble. But it is also an illustration of the brokering Wheatcroft is engaged in, picking up courtly or university poems from printed anthologies, and recycling them in his own 'book'.[11]

Much work remains to be done to see how the poetic of this mid-century yeoman developed under the influence of 'wit' as printed in various popular forms, but it is already apparent that 'Cum you galants' developed in close imitation of printed sources. What is more, it may be possible to see the actual development of the individual voice out of his selective copying. For example, no fewer than eight poems in the first 19 pages of 'Cum you galants' derive from one of the editions of the very popular mid-century anthology *Wits Recreations*, which Wheatcroft may have picked up, or seen, as a young man.[12] What is more, the general organisation of 'Cum you galants', consisting of a long section of mainly amorous verses followed by a last section of funerary elegies and epitaphs, follows one of the general principles of organisation in *Wits Recreation*. Wheatcroft's social and self-representational writings are undoubtedly partly formed by the example of popular printings, and in this context it is also not so surprising to find a courtship narrative attached to an anthology: part of the practical use of amatory verses and exemplary letters, at least in the earlier years, was in the business of courtship. The courtship narrative itself proves, as it were, the practical success of all this work of self-imaging, and if it worked for himself, then it might do also for others.

Of particular note, as I have already mentioned, is the ending of the narrative. Wheatcroft loved to be the centre of social attention, and the management of the wedding day and week or so afterwards probably rated as the best event in his life.

> The wedding days did last long. For eleven days together there was 11 dinners got. All was shot dinners; and there was which breakfasted, dined, and supped to the value of 200 persons, and I had one cook or two all the while. (p. 83)

He describes leaving Ashover early in the morning, sent off by neighbours and with nominated bellringers in action; how he arrived in Winster and they breakfasted, before the marriage at a justice's house; then how he had arranged a horserace back from Winster to Ashover across moor and dale with prizes of different coloured ribbons. He tells us who won, then how he had arranged a quintain at Ashover, for more provincial-chivalric sport. But his mastership of ceremonies excelled itself in supplying shot dinners, that is, meals which people came and paid for, for a period of eleven days – this was the equivalent of setting up house on the basis of wedding gifts. It was a long time to keep it going, drawing in families from far and wide. Then he has a typical ending, in which he signs off his pilgrimage, in lines of diminishing length, by calculating all the miles he covered to bring the campaign off:

... and to tell you how many times I went a-wooing you shall find so many slashes upon an ash tree at Winster town end, and how many miles I travelled for her sake was 440 and odd; and so gentle reader, I rest, Ashover Aug. 16 1657.

(pp. 88–9)

A calculation of investment validates romantic enterprise. Lest we wholly patronise this as the fall of chivalry into the hands of one of the lower middling sort, we should also remember that the paradoxes of romantic love, investment and risk inform many texts, not the least *The Merchant of Venice*.

Between these personal records from a provincial writer, we have differences. On the one hand, we have a display volume 'Cum you galants', which shows many of the characteristics of all those printed miscellanies which purveyed improving modes of writing to those of the middling sort, but there is within the miscellany a whole lot of personal material and most notably a whole narrative of his own courtship and marriage. On the other hand, we have an account of his own life, but that also includes other, non-personal materials evidently thought worth keeping and passing on. What is more, in the first book Wheatcroft moves texts from the miscellany to the courtship narrative, thus narrowing the gap between what is for entertaining display and what is personal record. And, just as there are non-personal intrusions into the autobiography, so some diversions in the courtship narrative can be large: in the first part, describing the journey to London, Wheatcroft does not pass up the opportunity to repeat at some length the subject of a sermon he heard at Barnet:

But what happened in all this way is too large to express, but a word or two I cannot let pass, which was of a sermon I heard at Barnet.

The text was *Believe on the Lord Jesus Christ and thou shalt be saved* (Luke 16:31). The words were answered of Paul and Silas to the gaoler ... [There follow several paragraphs of summary of the sermon.] Therefore I may well conclude that one drop of his most precious blood is worth ten thousand worlds.

This was taken notice of by me for the good of our immortal and never-dying souls.

The next day we advanced forth again, and about 9 of the clock entered into London, where I found my brother, who was no little glad to see me ... (pp. 38–9)

This sort of thing is not uncommon in accounts of journeys, but the sermon manifestly has nothing to do with courtship.

Then again, the readership of the autobiographical History might be thought to be more intimate or family-based than that of the courtship narrative, yet Wheatcroft happily cross-refers his books, as if readers were much the same:

> And in 55 I went awooing to my wife and was above two years before I was married to her and to tell you what hapned in our wooing I shall not heare, for you may find it at large in my book of Mirth and Melody. (p. 84)

Both narratives, being pilgrimages, are characterised by aspiration and achievement, and those qualities are culturally as well as personally defined. All in all, Wheatcroft presents an amusing case of what we all know, that kinds of personal record can never be free of elements of self-presentation, role-playing to a society, and in Wheatcroft's case, autobiographical recording is completely tangled up in a kind of laureate display.

Personal memorabilia are also, unsurprisingly, caught up in other kinds of recording of those things which were deemed worthy to treasure. That is to say, the beginnings of the cultural practice of writing personal records in the late middle ages and first parts of the early modern period, within annals of other more general kind, are clearly shown in Wheatcroft's texts: the most basic instinct is to *record* things thought worthy of memory, and the precise relation to the personal life can sometimes be less basic than that general impulse.

It must also be that we, as late twentieth-century readers, view this unstable mixture with degrees of prejudice. Perhaps one should not be surprised that personal recording and other kinds of record keeping are mixed up. For all his moments of pretentiousness, Wheatcroft is not in any case a sophisticated controller of form or genre. Yet if, for example, we find the conundrums meant for entertainment – 'Why have men beardes and women none' – particularly intrusive in the autobiography, that can only be because we wish, consciously or unconsciously, to read one thing above all in any 'autobiographical' account – revelations of what people 'really' did and thought. We seek the illusion of entering the unguarded mind. It is one of *our* common cultural interests, manifested in the continuing flood of biographies in recent years, the desire to 'get inside' the biographical subject.

On the other hand, so pervasive have been the reservations in critical and theoretical circles about the possibilities of reading an author as an individual, and so much have the possibilities been canvassed not of the personal presence but the author-effect, that autobiographical narratives are bound to be seen in those quarters as culturally constructed and as social negotiations in themselves. Indeed most of this writing, especially in the courtship narrative, is manifestly self-presentational. In the most interesting, if naive, fashion, Wheatcroft's blurring of kinds and his mixture of personal record and social display feeds precisely into the divided attitudes of a late twentieth-century scholarly readership.

With all that in mind, one might tell the further story of the self-presentational triumphs of Leonard Wheatcroft's death. As sexton, he had discovered in Ashover churchyard a medieval stone coffin. Here was a chance for immortality even better than his muses could afford. He ordered that his remains be placed into this monument of antique nobility, and buried. He had written a verse ready for the occasion:

> My will I'le make, and then Fare-well,
> Come louing saxton, Toule my Bell.
> In my Stone Coffin bury mee,
> That all my friends may plainely see,
> And at my head, A stone pray set,
> That where I ly, none may for get,
> I'ue writ ont so, he who goes by,
> Must stoupe to death as well as I,
> Death toules the bell / and strikes the Dart
> Fare-well deare friends, we all must part.
> Finis
> ('Galants', second pagination)

They discovered him and dug him up in 1880. They discovered also that someone had sought to secure for him further immortality. Inside the coffin a heart-shaped flat piece of lead had been placed on the body, on which was inscribed

HERE WAS LEONARD WHEATCROFT BURIED
JANUARY III. IN THIS STONE
COFFIN; WHO WAS CLARK OF THIS
CHURCH 36 YEARS
AGED 80
1706.[13]

A man can write his own epitaph, but not an inscription giving the precise date of death, although Wheatcroft got as close as he could to such an arrangement. A piece of paper at the beginning of the autobiography reads 'A History of my Birth, Parantage / And Pilgrimage who was / Borne May the first 1627 / and was Buried . . .', leaving someone else, presumably his son Titus, to complete the text (as he duly did). Perhaps Titus executed the last noble order of the heart-shaped piece of lead as well. You could *almost* control an audience which was your son, at least at the time of your death.

Notes

1 *Diary of John Evelyn*, ed. E.S. De Beer, vol. 4 (Oxford, 1955), p. 222, quoted in the introduction to *English Family Life: An Anthology from Diaries*, ed. Ralph Houlbrooke (Oxford and New York, 1988).

2 Both are deposited in the Derbyshire Record Office at Matlock. The autobiography is DRO, D.1101; the courtship narrative in the miscellany book is PZ 5/1 in DRO, D.253. The autobiography has most recently been edited by Dorothy Riden in *A Seventeenth-Century Scarsdale Miscellany* (Derbyshire Record Society Vol. XX, 1993) pp. 71–117. From this edition quotations are given. The old edition by Charles Kerry in *The Journal of the Derbyshire Archaeological and Natural History Society*, 21 (1899) 26–60, was full of inaccuracies. The courtship narrative has been edited by George Parfitt and Ralph Houlbrooke in *The Courtship Narrative of Leonard Wheatcroft: Derbyshire Yeoman* (Reading, 1986), and from this edition quotations are given. Excerpts from Wheatcroft's writings were given in Charles Kerry, 'Leonard Wheatcroft, of Ashover' in *The Journal of the Derbyshire Archaeological and Natural History Society*, 18 (1896) 29–80, and in Henry Kirke, 'Some Notes on the Minor Poets of Derbyshire', ibid., 44 (1922) 1–22. See also S.O. Addy, 'Ashover and the Wheatcrofts,' ibid., 34 (1917) 109–53, C. E. Lugard, *The Saints and Sinners and the Inns and Outs of Ashover* (Leicester, 1924, limited reprint, 1972), and W. Notestein, 'Leonard Wheatcroft, 1627 to 1706' in *English Folk. A Book of Characters* (London, 1938) pp. 205–14.

3 Quotations of verse are taken straight from the manuscript book, except where they occur within the edition of the courtship narrative. The book has two paginations, but some of the pages are so damaged that citations of individual page numbers are often not possible.

4 Both the popular naming of bards and the possibilities of public competition evidently occurred elsewhere: cf. the contest set up with a sale of tickets between John Taylor, 'The Water Poet', and William Fenner, 'The King's Rhyming Poet', in London in 1615. See *The Works of John Taylor, The Water Poet*, ed. C. Hindley (London, 1872) viii–x.

5 His son Titus' memoranda are at PZ 5/2 in D.253. See also C. Kerry, 'Ashover. Memoranda by Titus Wheatcroft, A.D. 1722,' in *The Journal of the Derbyshire Archaeological and Natural History Society*, 19 (1897) 24–52.

6 See Autobiography, p. 90 and p. 87.
7 See Autobiography, p. 90. Wheatcroft also records, as one of his notable doings, that he made an account book (p. 87).
8 There must be a possibility that his many epitaphs were a source of income, too. A man who was village sexton and church clerk was at the centre of funerary arrangements.
9 John Taylor, 'The Water Poet', provides another comparison at this point: a high number of his works was based on voyages and journeys, and all were designed to sell. The first-person narrator could evidently be used in such texts for effects of authentic witness, something which is also true of very many 'annals' of notable events, as for example George Wither's 'History of the Pestilence', his manuscript advice poem to Charles I (see the edition of J. Milton French (Cambridge, Mass, 1932), enlarged and printed as *Britain's Remembrancer* (London, 1628).
10 It appears in *Parnassus Biceps, Or Severall Choice Pieces of Poetry, composed by the best wits that were in both the Universities before their Dissolution* [compiled by Abraham Wright], (London, 1656) pp. 43–4, as the second half of 'Upon his chast Mistresse'. This miscellany dates from the same year as the culmination of the courtship, but no other poems are in common, so the means of transmission must remain doubtful.
11 Some of the discussion of the movement of texts from manuscript to print and print to manuscript can be found in Arthur F. Marotti, *Manuscript, Print, and the English Renaissance Lyric* (Ithaca and London, 1995).
12 The correspondences are with the first edition: *Wits Recreations. Selected from the finest Fancies of Moderne Muses* . . . (London, 1640). Some groups of poems are in the same sequence, suggesting a strong connection whether direct or indirect. My thanks to Adam Smyth for help with the printed miscellanies.
13 I have corrected the '56' years of the transcript in Kerry, *DAJ*, 18 (1896) 46, in accordance with the two other witnesses given on the same page.

9

They Only Lived Twice: Public and Private Selfhood in the Autobiographies of Anne, Lady Halkett and Colonel Joseph Bampfield

Sheila Ottway

The voice of vindication is one that is often to be heard in autobiography; indeed, a desire for personal vindication has traditionally been one of the prime motives for writing the story of one's own life. In this essay I describe how two seventeenth-century English autobiographers, Anne, Lady Halkett and Colonel Joseph Bampfield, came to write their own life stories with a view to defending their personal reputations, in public and in private; in my comparison and analysis of these two texts I explore the way in which gendered subjectivity may play a role in the articulation of self-definition and exculpation. In other words, I shall be considering what it meant, in seventeenth-century England, to be either a man or a woman whose good name was under threat, and how a sense of gendered identity can influence the form and content of an autobiography written with a view to self-justification.

Anne, Lady Halkett was born in 1623 (her maiden name was Murray; for the sake of simplicity I shall refer to her by her married name). As a young unmarried woman she was involved in helping the Royalist cause in England during the politically turbulent period of the 1640s and 1650s. She is remembered in particular for the part she played in helping James, Duke of York, the younger son of King Charles I, to escape from London to the Low Countries. This was in 1648, when the King was already a prisoner of the Revolutionary regime in England, under Cromwell. Anne Halkett was responsible for arranging a female disguise for the young duke, then aged 14, who was able to escape from a game of hide-and-seek in the gardens of St James's Palace, and to board a ship waiting for him on the Thames. In this cloak-and-dagger exploit, the duke was assisted and accompanied personally by both Anne Halkett and the Royalist secret agent, Colonel Joseph Bampfield. While all this political intrigue was going on, Anne Halkett and Joseph Bampfield became closely involved with each other not only professionally but also personally: in short, they had a passionate love affair. Soon after the successful abduction of the Duke of York, Bampfield came to visit Anne Halkett privately, and told her that he had received the news of the death of his wife; believing himself to be a widower, he asked Anne to marry him, and she consented, at least provisionally. But before the relationship could develop any further, it transpired that Bampfield's wife was perhaps not dead after all; Bampfield had apparently been wilfully deceitful. Anne Halkett and Joseph Bampfield subsequently went their different ways, but kept in contact with each other. Both continued to support the Royalist cause, in Scotland and in England. In 1654, they met up again in London. By this time, Anne had received an offer of marriage from a Scottish gentleman, Sir James Halkett. During this meeting in London, when Bampfield asked Anne if she were already married to this gentleman, it was now her turn to be deceitful, and she replied in the affirmative, whereupon the disappointed colonel went away and left her forever. Some 15 months later, Anne did indeed marry Sir James Halkett, and went to live with him on his estate in Scotland.

Many years later, Anne Halkett wrote her memoirs; indeed, it is thanks to this autobiographical record of the events of her life that we know so much about her. By the time Anne Halkett came to write her memoirs, in 1677 or 1678, she was a widow, looking back on the events of her life, in particular on the tumultuous years of the Interregnum. Her memoirs constitute a private document, never intended for publication.[1] The memoirs were published for the first time in 1875; a modern edition appeared in 1979.[2] This secular autobiography is remarkable for its narrative pace and stylistic artistry; as one critic has remarked, 'Filled with romance, excitement and suspense, Halkett's *Autobiography* exhibits the skills of a psychological novelist.'[3] Indeed, various critics have compared Anne Halkett's memoirs to such classic novels as Samuel Richardson's *Clarissa*, Jane Austen's *Sense and Sensibility* and Charlotte Brontë's *Jane Eyre*.[4] (My somewhat fanciful title for this essay suggests that the twentieth-century spy novel could be added, if somewhat bathetically, to this illustrious list.) Anne Halkett's autobiography focuses on her participation in the events of the political struggle for power in Britain during the 1640s and 1650s, and on the three courtships of her life, the second of which involved Joseph Bampfield. The text is thus concerned with both public and private life, which for Anne Halkett were closely interwoven. The fact that she chose to record the events of her life in a private document can be seen as a consequence of prevailing attitudes towards women's conduct at the time when Halkett was writing her autobiography in the 1670s: after a period of relative freedom of expression for women in England during the politically unstable Interregnum, there followed a reactionary backlash against this trend in the male-dominated cultural climate of the Restoration. If Anne Halkett had published her autobiography at that time, it would surely have met with public censure. It seems most likely, therefore, that she wrote the story of her life for the exclusive readership of her own immediate family.

Various critics have praised Anne Halkett's autobiography for its narrative impetus and stylistic sophistication; in the words of Paul Delany, for example, 'we should give Lady Halkett credit for being one of the most perceptive and skilful stylists among British autobiographers of her time'.[5] A more recent critic, N.H. Keeble, has said of Halkett's autobiography that 'it is the resilience with which she engages in a hostile world and the ingenuity by which she surmounts sudden hazards in which the narrative delights'.[6] Halkett's autobiography is indeed a celebration of the independence and self-determination of its protagonist; according to N.H. Keeble its author certainly succeeds in fashioning 'a female self who is anything but an obedient feminine subject'.[7] While agreeing with these views, I would suggest that Halkett's text is in fact both celebratory and confessional: in particular it is Halkett's love affair with Joseph Bampfield that forms the thread of her narrative, and it is for her involvement in this affair that she seeks personal vindication. Halkett consistently defends herself throughout her autobiography in her dealings with Bampfield, emphasising how she had trusted him because of his apparent sincerity. In her own words:

> when ever hee came, his discourse was serious, handsome, and tending to imprese the advantages of piety, loyalty, and vertue; and these subjects were so agreeable to my owne inclination that I could nott butt give them a good reception, especially from one that seemed to bee so much an owner of them himselfe.[8]

When Bampfield tells her that his wife is dead, she has such faith in his integrity that she is willing to believe this for a long time, until the truth of the situation emerges, and Bampfield's wife turns out to be very much alive. Nevertheless, during the long period of uncertainty as to Bampfield's marital status, Anne Halkett clearly had a deeply committed relationship with him, one of trust and intimacy. It is surely because of the details of this intimate affair that two leaves of the manuscript of Halkett's autobiography were intentionally torn out by some unknown person, presumably in order to protect the author's reputation. These two tantalising gaps in the text make one wonder precisely what had been said in the missing pages, particularly as they both occur in passages dealing with Anne Halkett's relationship with Bampfield.[9] The second of these gaps is especially suggestive, as it occurs at a point in Anne Halkett's narrative where she recalls a delicate conversation between herself and her future husband, Sir James Halkett, in which he informs her that Bampfield's wife is obviously alive:

Att last hee said, 'I have heard news this morning, and though I know itt will trouble you, yett I think itt is fitt you should bee acquainted with itt. Just as I was turning down Blacke-fryar Wind (said hee) to come here, Collonel Hay called to mee and told mee the post that came in yesterday morning had brought letters from London that undouptedly C.B.'s [Colonel Bampfield's] wife was living and was now att London, where shee came cheefely to undeceave those who beleeved her dead.' 'Oh,' said I, with a sad sigh, 'is my misfortune so soone devulged? . . .' [one leaf (two pages) missing] . . . unworthy, and in what apeared so, none living could condemne mee more than I did my selfe.[10]

What precisely was the 'misfortune' thus 'devulged' that Anne Halkett refers to here? Had she lost her virginity, or had she perhaps been married bigamously, though unwittingly so, to Joseph Bampfield? This is indeed suggested in a later passage in the text, when Anne Halkett, obviously distraught, seeks advice from a clergyman, a Mr Dickson. Fortunately, he is able to assure her that, in her own words:

since what I did was suposing C.B. a free person, hee nott proving so, though I had been puplickely maried to him and avowedly lived with him as his wife, yett the ground of itt failing, I was as free as if I had never seene him; and this, hee [the clergyman] assured mee, I might rely upon, that I might without offence either to the laws of God or man marry any other person whenever I found itt convenient . . .[11]

As a result of this interview, Anne Halkett feels herself to be relieved of her commitment to Bampfield, whom she therefore feels justified in rejecting at their next and final meeting. Thereafter she is able to accept Sir James Halkett's proposal of marriage with a clear conscience, it would seem, with good prospects of living happily ever after. And yet, there are many loose ends to this story, not least the fact that the narrative breaks off suddenly after the account of Anne's marriage to Sir James Halkett in 1656. How many more pages are missing, one wonders?

Whatever the answer to this question might be, it is clear that the autobiography as it stands is dominated by Anne Halkett's recollections of her relationship with Joseph Bampfield. It would seem that the motivation for writing her autobiography was partly to celebrate her public involvement in the Royalist struggle for power, and partly to vindicate herself from any possible slander concerning her private love affair with Bampfield. According to John Loftis, the editor of the 1979 edition of Halkett's text, 'Lady Halkett was still deeply troubled in conscience about her relationship with Bampfield at the time she wrote her memoirs.'[12] Loftis speaks of 'the tension she experienced between passionate affection and, in opposition to it, a habitual moral rectitude reinforced by religious conviction'.[13] Moreover, Loftis clearly declares his partiality for Anne Halkett when he refers to Bampfield as 'an unconscionable opportunist', stating categorically that 'we feel resentment at him, despite the three intervening centuries'.[14] Such, it would seem, is the remarkably emotive power of the voice of a damsel in distress.

If Anne Halkett was able to save her own reputation by writing her autobiography, by doing so she succeeded in giving Joseph Bampfield an unequivocally bad name. He has gone down in history not only as a man who was duplicitous in his personal relationships with women, but also as a man who played the devious role of a double agent in his professional capacity in the world of political espionage.[15] Fortunately, however, he too wrote an autobiography, entitled *Colonel Joseph Bampfield's Apology*, published in 1685; a modern edition of this text appeared in 1993. In order to explain why he wrote this autobiography, I must briefly describe the course of his career.[16] Born most probably in 1622 or 1623, Joseph Bampfield joined King Charles I's army at the age of 17. He soon rose through the ranks, and became the colonel of a regiment in the early stages of the English Civil War. As the war progressed, Bampfield became employed as an intelligence agent by Charles I in various secret negotiations; as mentioned earlier, he was a key figure in organising, with the aid of Anne Halkett, the escape of the Duke of York from London in 1648. Clearly, Bampfield was at this time utterly devoted to the Royalist cause and did all that was in his power to help save Charles I from being deposed by the Revolutionary regime. This, alas, was his undoing, for at some time during the desperate few months preceding the execution of the King, in 1649, Bampfield wrote a fateful letter to Charles I, which came to be read by Charles, Prince of Wales (later to become Charles II), who was at that time living in exile in the Low Countries. In this letter, Bampfield had suggested that a rescue operation could be organised to release the King from his imprisonment on the Isle of Wight, and that, in Bampfield's own words, 'it would be a great security to his affairs [that is to say, those of Charles I], and of no less encouragement to his friends, when his children were at the head of those who should appear for him.'[17] In other words, Bampfield was suggesting that the Prince of Wales should be at the head of an invading force from the Continent that would storm Carisbrooke Castle on the Isle of Wight and carry the King away to safety. By this time, however, things were looking pretty bleak for the Royalists; earlier that year (1648), the Prince of Wales, when in command of the Royalist fleet, had already failed to make any attempt to rescue his father, at a time when this might well have succeeded. It seems that the Prince of Wales was greatly offended when he read Bampfield's letter, presumably because of his guilty conscience; Charles did not like to be reminded by a mere underling that he had not tried to rescue his imprisoned father when this might have been possible, and now it was apparently too late. In any event, Bampfield fell from royal favour, which made his career as an intelligence agent extremely difficult. Who could he work for after this unintended blunder, except for another political regime?

It is therefore not surprising that the disillusioned Bampfield eventually sought and found employment in the rival camp, Cromwell's Protectorate, in 1654. Bampfield served his new masters until the Restoration of the monarchy in 1660; then, of course, he became once again a *persona non grata*. Unable to stay in England, he fled to the Dutch Republic, where he was received into the service of the state. Here he was employed for several years by the statesman Johan de Witt. After 1672, known as the Dutch *rampjaar*, or year of disaster, following an unfortunate military debacle in an engagement with the French, Bampfield went to live in retirement in the far north of the Netherlands, in the province of Friesland. Here he kept a low profile, living quietly, for some time in the small town of Bergum, and later in the city of Leeuwarden. It was in these years of his retirement that he evidently wrote his memoirs. It seems that he was prompted to publish them, in the form of an apology or defence, when the discovery was made in England in 1683 of a conspiracy (the Rye House plot) against Charles II, Bampfield's old enemy. The exposure of this plot provided the King with the opportunity to prosecute supposed enemies of the state, and if necessary to secure the extradition of such persons from abroad. As his autobiography suggests, Bampfield was at this time a deeply worried man, living in fear for his life. Would Charles's own spies be coming to get him? The possibility must have seemed all too real at the time.

Unlike Anne Halkett's autobiography, Bampfield's *Apology* was written expressly for publication: on the title-page it is stated that this text has been 'written by himself and printed at his desire', suggesting that he actually had it published at his own expense. In fact, Bampfield saw to it that his *Apology* was published simultaneously in both English and Dutch.[18] Bampfield's autobiography is an apology in the strict sense of the word. That is to say, it is not an expression of regret or remorse for past actions, but rather an assertion of personal integrity, with a view to vindicating the author's public reputation. In his *Apology*, Bampfield relates the story of his professional career as a soldier and intelligence agent; it is thus essentially a factual document, with detailed descriptions of military operations and secret missions in which he had been involved. It is written in an orderly fashion, in numbered paragraphs, with a preface explaining why the author has written this account of his life: namely, in order to counter false accusations that have been made against him, and to establish his innocence. In his preface Bampfield speaks of the 'most insupportable injuries and calumnies of some who are totally ignorant of the truth of my affairs', and of 'the cruelty of their secret practices and obscure proceedings against me to the endangering of my life'.[19] Bampfield stresses that it is his masculine honour that is at stake, stating that:

> This being really my case, [it] is also the sole cause why I am enforced to expose my discretion to the capricious censure of critics, rather than to abandon my honor by a womanish modesty, timidity, and silence...[20]

In his account of his life and career, Bampfield faces the difficult task of explaining why he changed his political allegiance. He attempts to deny his reputation as a renegade or turncoat, but he can only do so by stressing that his changes in allegiance were enforced by circumstance, notably by the unreasonable enmity towards him on the part of Charles II. Clearly, Bampfield had felt betrayed by Charles, especially since he had so dutifully served Charles's father, Charles I, and his younger brother, the Duke of York, at the risk of his own life.

Bampfield's *Apology* is therefore not simply a chronicle of one man's part in the political events of seventeenth-century England; it is at the same time an expression of personal vindication, and, on occasion, a veritable cry from the heart. Bampfield's state of dejection in Friesland in the 1680s is painfully evident, to say the least; thus, in his preface he speaks of 'this sepulcher wherein I have for some years lain buried alive'.[21] And in recalling the time when he was living in Bergum, he describes his situation there as 'this Egyptian darkness': a poignant metaphor alluding to the biblical account of the exile of the Israelites in Egypt.[22] In his own lonely exile in Friesland, Bampfield must have felt like a spy who had been literally left out in the cold. But among the various metaphorical expressions that Bampfield uses to add vividness to his narrative, the most memorable are those relating to shipwreck. (Perhaps this is appropriate for a man who must have spent a lot of his time travelling on secret missions to and fro across the North Sea.) Thus, in his *Apology*, after briefly describing how he had accomplished the dramatic rescue of the Duke of York in 1648, Bampfield says, concerning the young duke and his followers, that 'the wind, which was favorable to bring them into Holland, proved a storm to me, which occasioned my wrack ever since, having unhappily given some counsel concerning that fleet (it seems) with too much precipitation, which was well meant, [but] very ill taken, because most bitterly represented by a person of quality' (and here of course he means Charles II).[23] And in another passage, Bampfield relates how after he had fallen into Charles II's disfavour he was rejected by his former associates, whom he refers to as those who 'abandoned me as perishing men in a shipwreck save themselves as they can'.[24] And finally, in describing his present situation in Friesland, he speaks ominously of 'this dead calm' in which he finds himself: an apt description of his evident state of torpor in the face of possibly imminent death. It is surely by means of these maritime metaphors that Bampfield attempts to evoke the sympathy of his readers.

The question arises as to whether Bampfield's defence of his reputation in his *Apology* is convincing. Interestingly, it would seem that his reputation has been redeemed with the recent republication of this text, co-edited by John Loftis. As mentioned previously, Loftis was also the editor of Anne Halkett's *Memoirs* in 1979, referring to Bampfield in no uncertain terms in his introduction as an 'unconscionable opportunist'. But in the editors' preface to Bampfield's *Apology*, we find a very different view expressed: here it is said that

John Loftis must add a defensive note to explain the difference in his attitude toward Bampfield expressed here from that apparent in his edition of *The Memoirs of Anne, Lady Halkett*... of 1979. The severity of his remarks about Bampfield in the earlier book, not inconsistent with most twentieth-century comment on him, were in part the result of admiration for Lady Halkett and uncritical acceptance of her attitudes. One reviewer of the edition noted that Loftis had 'a soft spot for Lady Halkett'; true – and a corresponding hostility toward the man who, he then thought, fabricated a report of his wife's death as a means to seduce her.[25]

So what has happened here? It turns out that Loftis now sees Bampfield as a man of unimpeachable integrity who was betrayed in both public and private life, with dire consequences. As for Bampfield's public reputation, Loftis has reassessed his opinion in the light of modern historical studies and certain seventeenth-century documents published only this century, sources which indicate that Bampfield's record of service in the Dutch republic was incompatible with the

> conception of the man he [that is, John Loftis] had formed when editing the earlier book. Dutch references to Bampfield, free of the distorting passions embittered by the English civil wars, reveal him as a mature man of good will and good judgment, acting responsibly in positions of trust.[26]

Moreover, Loftis and his co-editor Paul Hardacre refute the evidently 'erroneous assumption' that Bampfield had ever acted in England as a double agent.[27] And as for Bampfield's private life, here too he is now exonerated from duplicity: according to Loftis, it is clear that Bampfield had been wilfully misinformed of the death of his wife, Catherine, by an unfaithful servant. Apparently, Bampfield was acting in good faith when he proposed marriage to Anne Halkett, thinking himself to be a widower; only later did he come to find out the truth.[28] Once again he had been betray In exposing Bampfield's victimisation through betrayal, Loftis thus makes a strong case for his historical rehabilitation, finding support for this view in his compassionate reading of Bampfield's autobiography.

Nowhere in his *Apology* does Bampfield mention the name of Anne Halkett. This emphasises one of the essential differences between the two autobiographies discussed here: namely, that in Western patriarchal society, vindication for a man has traditionally had more to do with his public image, his sense of honour, and his devotion to some form of secular authority, whereas for a woman, vindication has had more to do with her private relationships and her sexuality.[29] In early modern England, a man's reputation generally depended on how he conducted himself in public affairs, while a woman's reputation depended on her chastity. Both Joseph Bampfield and Anne Halkett failed, in a sense, to meet the requirements made of them by the dominant culture of their time; nevertheless, both of them succeeded in articulating self-justification in their autobiographies, by emphasising the mitigating circumstances that encouraged them to conduct their lives in the way they did. As mentioned previously, I have given this essay a somewhat facetious title; in doing so, however, it has not been my intention simply to compare the life stories of two seventeenth-century autobio-graphers with contemporary spy fiction. Rather, I would point out the fact that we all, in a sense, lead double lives: a public and a private life, which we may keep carefully separate or which we may allow to become closely interwoven. The autobiographer, moreover, has a double life of a different kind: the life that is lived in reality, and the life that is inscribed for posterity. In this essay I have attempted to give Joseph Bampfield and Anne Halkett an extra lease of life together, if only metaphorically, by briefly retelling their stories. At the same time, I hope to have demonstrated the remarkably persuasive power of auto-biography in the making or remaking of an individual's historical reputation.

There is a final chapter to this story, or perhaps I should call it a postscript. In 1701, two years after the death of Anne Halkett, a bio-graphy was published, entitled *The Life of the Lady Halket*, written by a mysterious author referring to himself simply by his initials, 'S.C.' (I assume this was a man because the author refers to himself as 'he' in the preface.)[30] The text is obviously based on Anne Halkett's own memoirs, to which, it would seem, the author had access before the two missing leaves were removed. This is suggested by the fact that the author refers to Anne Halkett as having been at some time in Holland – something which is never mentioned in her autobiography as we know it. It seems very well possible, as John Loftis suggests, that Anne Halkett and Joseph Bampfield had travelled to the Netherlands together to be married, and that this is precisely what was recorded on the pages of the manuscript of her memoirs that were removed and presumably destroy.[31] Interest-ingly, the shipwreck metaphor appears now, in this biography, with reference to the life of Anne Halkett; the author thus speaks of 'this excellent lady ... being tossed, as it were between waves and pursued with a constant series of difficulties and encumbrances for the space of fourteen years, both in England and Holland; till at length the one ship-wrack'd and bereaved of all comforts (except her vertue and integrity) she arrived at some settled state.'[32] Nowhere in her memoirs does Anne Halkett refer to herself metaphorically in this way. One wonders whether the mysterious Mr S.C. had read Bampfield's *Apology*, and was intentionally reacting to it, by using the same kind of metaphor for the misfortunes of Anne Halkett as Bampfield had done for his own. In any case, Mr S.C. clearly portrays Anne Halkett as a woman who overcame her tribulations and who was exceptionally devout, especially in later life. At the end of the biography a list is given of the contents of the 21 devotional books that she wrote. Her biographer also emphasises Anne Halkett's skill in 'the study of Physick', and in the preparation of 'such medicines, and drugs as might be helpful in common and ordinary diseases'.[33] Moreover, the author stresses Anne Halkett's piety by appending to his biography the text of her meditation on the 25th Psalm, a psalm which, perhaps significantly, includes the following words, in verse 7: 'Remember not the sins of my youth, nor my transgressions; but according to thy mercy remember me, for thy Goodness sake, O Lord.' One wonders whether Mr S.C. thought that Anne Halkett's vindication needed a little scriptural underpinning.

Notes

1 The manuscript is now in the British Library: Manuscript of Lady Halkett's Memoirs, B.L. Add. MS. 32376.
2 See *The Autobiography of Anne, Lady Halkett*, ed. John Gough Nichols (London, 1875); and *The Memoirs of Anne, Lady Halkett and Ann, Lady Fanshawe*, ed. John Loftis (Oxford, 1979). I have used the latter edition for the purposes of this essay.
3 Estelle C. Jelinek, *The Tradition of Women's Autobiography: From Antiquity to the Present* (Boston, 1986) p. 31.
4 See, respectively, James Sutherland, *English Literature of the Late Seventeenth Century* (Oxford, 1969) p. 263; Margaret Bottrall, *Every Man a Phoenix: Studies in Seventeenth-century Autobiography* (London, 1958) p. 153; and Donald A. Stauffer, *English Biography before 1700* (Cambridge, 1930) p. 263.
5 Paul Delany, *British Autobiography in the Seventeenth Century* (London, 1969) p. 164.
6 N.H. Keeble, 'Obedient Subjects? The Loyal Self in some Later Seventeenth-century Royalist Women's Memoirs', in *Culture and Society in the Stuart Restoration: Literature, Drama, History*, ed. Gerald Maclean (Cambridge, 1995) pp. 201–18, especially p. 211.
7 Keeble, 'Obedient Subjects?', p. 215. Here Keeble is referring to the memoirs not only of Anne, Lady Halkett, but also those of Ann, Lady Fanshawe and Margaret Cavendish, Duchess of Newcastle.

8   *The Memoirs of Anne, Lady Halkett*, p. 23.
9   In John Loftis's edition of Anne Halkett's memoirs the gaps are indicated on pp. 23 and 72.
10 *The Memoirs of Anne, Lady Halkett*, p. 72.
11 *The Memoirs of Anne, Lady Halkett*, p. 76.
12 John Loftis, 'Introduction' to the *Memoirs*, p. xiii.
13 John Loftis, 'Introduction' to the *Memoirs*, p. xi.
14 John Loftis, 'Introduction' to the *Memoirs*, p. xii.
15 In the *Dictionary of National Biography*, for example, it is said that 'he was justly suspected by Charles II to be acting a double part' in the struggle for power between the Commonwealth and the Crown during the 1650s. See the *DNB*, Vol. II, ed. Leslie Stephen (London, 1885) p. 102. For an assertion of Bampfield's duplicity in relations with women, see Antonia Fraser, *The Weaker Vessel: Women's Lot in Seventeenth-century England* (London, 1984) pp. 188–9, 204.
16 This information is derived from the editors' 'Introduction' to the 1993 edition of Bampfield's *Apology*, as well as from the survey of Bampfield's later career, a biographical supplement by John Loftis, included in the same volume. See *Colonel Joseph Bampfield's Apology: Written by Himself and Printed at his Desire*, ed. John Loftis and Paul H. Hardacre (Lewisburg, 1993). All page references are to this edition (hereafter abbreviated to *CJBA*).
17 *CJBA*, paragraph 115, p. 78.
18 The Dutch version is entitled *Apologie van Colonel Joseph Bampfield* (The United Provinces, 1685). For details of this Dutch text, see 'A note on the text', by the editors, John Loftis and Paul H. Hardacre, in their edition of Bampfield's *Apology*, pp. 35–6.
19 *CJBA*, paragraph 1, p. 37.
20 *CJBA*, paragraph 3, p. 38.
21 *CJBA*, paragraph 2, p. 38.
22 *CJBA*, paragraph 152, p. 92.
23 *CJBA*, paragraph 97, p. 71.
24 *CJBA*, paragraph 129, p. 83.
25 'Preface' to *CBJA*, p. 14.
26 'Preface' to *CJBA*, p. 14.
27 See 'Introduction' to *CJBA*, p. 21.
28 See John Loftis's 'Conclusion' in *CJBA*, pp. 245–6.
29 Stephen Greenblatt sees an individual's relationship with authority as one of the governing conditions for self-fashioning in the early modern period. See Stephen Greenblatt, *Renaissance Self-fashioning from More to Shakespeare* (Chicago, 1980) p. 9. Greenblatt's study of self-fashioning deals exclusively with male case studies; thus his universalising theories about the construction of the self are clearly put forward with only men in mind.
30 S.C., *The Life of the Lady Halket* (Edinburgh, 1701).
31 See John Loftis's 'Conclusion', in *CJBA*, p. 248.
32 *The Life of the Lady Halket*, p. 13.
33 *The Life of the Lady Halket*, p. 10.

10
[Re]constructing the Past: the Diametric Lives of Mary Rich

Ramona Wray

> We must recognise what the past suggests: women are well
> beyond youth when they begin, often unconsciously, to create
> another story. Not even then do they recognise it as another
> story. Usually they believe that the obvious reasons for what
> they are doing are the only ones; only in hindsight, or through
> a biographer's imaginative eyes, can the concealed story be
> surmised.[1]

From the amount of her extant work, it is evident that Mary Rich,
Countess of Warwick, devoted considerable time to the process of writing
and rewriting her own life.[2] Her diary runs to thousands of manuscript
pages, with almost daily entries beginning on 25 July 1666 and
continuing until eighteen months before her death in 1678.[3] Rich also
composed a short autobiography.[4] Only 40 pages in length, it dates
from around 1671 and recounts an entire personal history from birth to
old age. Both narratives have as their main thrust Rich's story and both
are ostensibly private, although traces of revision, a refusal to detail
particularly sensitive issues and signs of self-censorship may suggest
that, at some level, a reader was borne in mind.[5] The distinction
between the texts is additionally blurred since, at one point, at least,
Rich was writing them both simultaneously.[6]

These points of contact between autobiography and diary make
the contradictory and mutually destabilising nature of the two texts
particularly intriguing. Strikingly, while the two texts relate many of
the same experiences – early years spent in Ireland and London, a
controversial marriage to the impoverished Charles Rich, a religious
conversion a few years into the marriage, the death of two children and

widowhood – they disagree *vehemently* on the particular inflections that these experiences attract. In the diary, for instance, Rich's husband, Charles, is constructed as a violent and abusive tyrant, in contrast to the autobiography where he is shown as a gallant romantic hero. The Rich portrayed in the diary is chronically depressed, disappointed and embittered, a far cry from the lively, confident and fulfilled woman described in the autobiography. The diary relationship between Rich and her husband is characterised by conflict, disharmony and mental abuse, while in the autobiography the relationship is discovered as a loving ideal. In the autobiography and the diary, then, are two very different self-presentations, two lives, which are diametrically opposed and united only in their essential experiential contours.[7]

Clearly, the existence of these two contrasting self-representations calls for comment. One obvious question is why Rich chooses to write in such opposed modes, why she elects to write her life twice over. So far, the definitive discussion of the forces motivating women to write themselves has been provided by Mary Jane Moffat and Charlotte Painter. They argue that:

> dissatisfaction with the way love and work have been defined for the female is the unconscious impulse that prompts many [women] to acquire the habit of personal accounting on some more or less regular basis.[8]

On initial impressions, this theory of diary writing as a response to feelings of frustration and unhappiness strikes an immediate chord of recognition with a reading of Rich's texts. The diary creates a potent impression of repeated disappointments, deep dissatisfactions and continuous blows to a fragile self-esteem. The particular focus for this sense of discontent is Rich's troubled relationship with her husband. Despite the fact that the marriage between Charles and Rich was a love-match, contracted without family approval, the diary records numerous instances of marital disharmony:

> My lord did so provoke me that I was surprised into a sudden disputing with him ... I fell into a dispute with him ... After I returned home, I fell into a foolish dispute with my lord ... I fell into a dispute with him wherein I was very passionately affected. (*Diary*, pp. 95, 136, 118)

Rarely does Rich dwell upon the specificities of the altercation: it is the generality and pervasiveness of the friction that dominates above all else. The degree to which Rich uses the same phrasing and the same terminology when she articulates these moments of marital crisis means that her narrative takes on some of the properties of a balladic refrain:

> After dinner, without any occasion given, my lord fell into great passion with me... My lord was passionate with me without any occasion... He fell violently passionate against me... My lord, without any occasion given by me, fell into a great passion with me. (*Diary*, pp. 74, 131, 103, 136)

Comments such as 'his fierceness made me troubled that I had disputed with him' (*Diary*, p. 95), in conjunction with Rich's frequent resolutions not to quarrel with Charles for fear of the consequences, suggest that their disagreements may have gone beyond the verbal, and that Rich was the victim of domestic abuse.[9]

Faced with such evidence of Rich's unhappiness, it is tempting to argue, as Moffat and Painter might, that she turned to her diary as a refuge from everyday tensions and dissatisfactions. Certainly the diary suggests that writing often afforded Rich emotional release and represented, for her, something akin to modern therapeutic practice. On numerous occasions Rich describes the writing process in terms of release:

> My mind being still much disturbed, by reason of my lord's passion, I went to retire, and was enabled to pour out my soul, and to weep bitterly, and to lay [out] all my trouble as to a compassionate friend ... after this I found a great deal of ease. (*Diary*, pp. 154–5)

By means of her diary, Rich discloses the unspeakable, unveils what can never be shown to the outside world. The transaction, it seems, is one that brings comfort in the extreme.

However, if such passages suggest a possible impetus for writing, they do not fully answer to Rich's motivation. Crucially, a theory linking diary composition and quotidian unhappiness does not account for the *particular complexities* of Rich's self-representation – the very different configurations of Rich's experiences in the diary and autobiography. What particular needs are fulfilled by writing one's experience in opposed and often contradictory ways? By reading Rich's autobiography and diary in juxtaposition, this chapter suggests that Rich constructed a series of different autobiographical selves, not only as an outlet for tension through private expression, but, more interestingly, as a way of positively *altering* 'reality', both as she remembered it and as she experienced it. Vital to Rich's project, I argue, is the manner in which she shapes her raw material (the events of her life) to mime at least two literary paradigms, each of which is informed by a host of conventions. In their plot elements, character behaviours, language and values, Rich's two texts demonstrate a debt to the contemporary literary genres of romance and exemplary biography. The following two sections examine her appropriation of each literary form. Both sections begin by illuminating the particular borrowing process before going on to contend that both romance and exemplary biography appealed to Rich (in different and often contradictory ways) as a means of mediating a quotidian experience otherwise lacking in sources of emotional and spiritual sustenance. By appropriating the conventions of exemplary biography, Rich is empowered to transform her status from unwitting victim to tragic heroine, to find meaning in her suffering, to anticipate a happier psychological configuration and to take pragmatic steps towards bringing about changes in her present circumstances. By exploiting the conventions, logic and morality of romance, Rich is enabled both to justify the filial disobedience that any other discourse would castigate and to register how, through an unwitting investment in romantic ideology, she arrived at the poor choice that underpinned her marital decision. In conjunction, Rich's creation of a series of fictionalised selves enables her effectively to *rewrite* her past, present and future, thereby deriving from her restricted situation a greater satisfaction.

Such rewriting or modelling on Rich's part obviously required a familiarity with literary forms. Precisely what knowledge Rich would have had of romance and exemplary biography is inevitably a matter of speculation. The study of the gendered reception and ownership of texts is still in its infancy.[10] Several contemporary statements, however, establish that Rich was cognisant of popular literary genres. Rich's father, for instance, presented Mary with Sidney's *Arcadia* when she was 12.[11] And, both in the diary and in the autobiography, Rich records that, between approximately 1639 and 1650, she was 'spending my precious time in nothing else but reading romances, and in reading and seeing plays' (*Autobiography*, p. 8).[12] Furthermore, Rich refers in her diary to having spent a morning reading 'The Life of pious Mrs Smith' (*Diary*, p. 224). The title, and Rich's description, make it clear that the book is an exemplary biography similar to a host of others proliferating in the period.[13]

But my employment of these passing allusions is not to imply that Rich's deployment of romance and exemplary materials was necessarily part of a conscious or premeditated programme. Such fictions are not, of course, simply manipulated, but are simultaneously internalised as part of the way in which we experience our existence in the first place. The recollection of our lives is entangled with cultural models of experience (indeed, recollection is invariably conducted through such discursive mechanisms) to the extent that a mutually determining dialectic is often created. In this sense, the relationship between autobiography and fiction in Rich's writing is to be valued not only because it reveals the processes whereby a woman constructed her own life in the early modern period, but also because it allows us a glimpse into the way in which an individual female consciousness intersects with a prevailing cultural consciousness.

*

Nowhere is Rich's indebtedness to romance motifs and structures more apparent than in her textual realisation, in the autobiography, of her husband. Charles Rich, seen in the diary as a violent and abusive tyrant, appears in the autobiography as a hero taken straight from the pages of a romantic novella. In contrast to other male figures delineated in Rich's text, who are figured in terms of their innate characteristics, Charles is described wholly in relation to his appearance and demeanour. Despite his dominant role in the narrative, we learn little about him beyond the fact that he is 'a very cheerful and handsome, well-bred, and fashioned person' (*Autobiography*, p. 3). In elaborating only his surface qualities, Rich blanks out those negative traits for which he might be arraigned, emphasising instead attributes prioritised in contemporary constructions of the romantic hero.

In conjunction with the romantic elevation of Charles goes the comparably heightened idealisation of Rich herself. Rich constructs a portrait of herself as an eminently eligible heiress by remarking upon the 'many very great and considerable offers' of marriage proffered by persons 'of great birth and fortune' throughout her teenage years (*Autobiography*, p. 3). At no point is the dovetailing between Rich and the romantic heroine more evident than when she expresses her dissatisfaction with the prospect of an arranged marriage. Faced with 'many great and advantageous offers', she cannot 'endure to hear of any of them' and can grant her suitors no satisfaction (*Autobiography*, p. 4). True to the spirit of the high-minded aristocratic protagonist, Rich turns her back on convention to find her own destiny, even as her narrative simultaneously reveals its indebtedness to a panoply of determining stereotypes.

Given Rich's allocation of the roles of hero and heroine to Charles and herself respectively, it is not surprising that the bond between the two of them is depicted as a perfect love-match. In the autobiography are none of the blazing arguments and bitter domestic disputes which characterise their relationship in the diary. Instead, frequently punctuating the narrative are references to Charles' 'great passion', and these are invariably accompanied by the acknowledgement that he holds 'the great and full possession ... of [Rich's] heart' (*Autobiography*, p. 4). Such an idealised construction of the relationship can, as might be expected, only be sustained through an almost exclusive focus on the period of courtship. Easily translatable into an unmediated romantic form, the inception of Rich's relationship with Charles proves the most enticing material for a fictional scenario. Rich devotes nearly half of her text to the period before marriage and is notably silent on the subject of her relationship with Charles once they are man and wife. The autobiography thus relies on the weight of romantically shaped cultural expectations to imply that the lovers lived happily ever after.

Why did Rich elect to place her life in an autobiographical romantic frame? Paradoxical as it may sound, I would like to suggest that a key to unravelling her motivations can be found in the powerful *critique* of romantic love elaborated in the diary. The diary firmly registers the discrepancy between the idealisation of the heterosexual union in romance and the 'reality' of married life. The process is conducted in two ways. First, there is an implicit criticism of romantic aspirations in the discovery of the love-match's 'actuality'. A romantic relationship is not continued into the diary, pointing up the potentially transient nature of sentimental attachment. Secondly, criticism is manifested through the diary's trenchant attack on romance's delusive qualities and coercive power. At those moments in the diary when Rich meditates upon the theme of unrealised expectations and broken dreams, her construction in the autobiography of Charles as 'the source of all her happiness' (*Autobiography*, p. 9) attracts particularly ironic inflections. The common preoccupation of such moments is the gap separating Rich's hoped-for life and its material embodiment. The disparity between this polarity is encapsulated most powerfully in statements which find their *raison d'être* in arresting antithetical conjunctions; as Rich writes, her 'bitterest crosses come from that creature that I did expect my sweetest comforts from' (*Diary*, p. 219), and her distress emerges from 'my having over loved that endeared relation from which I now met with so much unkindness'.[14] Via its juxtaposition of oppositional possibilities, Rich's text draws heightened attention to her life's disappointments.

However, her meditations refuse to place the blame for these disappointments either upon on 'that creature', Charles, or upon material circumstances. Instead, Rich's unhappiness is accounted for in terms of her unreasonable and ridiculous ambitions and youthful priorities. Onto *her own romantic longings* are displaced her ill-founded affection and poor choice of marriage partner. Rich broods, for instance, on 'the great dissatisfactoriness I had found in [the] things that I had set my heart upon and expected happiness from', illuminating a profound disjunction between what she has been encouraged to believe by an insidious cultural mythology and what hard experience has taught her to confront. Her strategy for coping is to resolve that 'it was not safe to let my heart ever again too freely go out unto any person or things of this world' (*Diary*, p. 219). In such expressions, Rich comes close to the realisation that romantic love has been less an ideal for her sense of well-being than a chimera that has cost her personal content.

These diary entries, all dated some years before the autobiography was composed, indicate that, at some level, at least, Rich (*while* she was fictionalising her life in the autobiography) simultaneously recognised the constructedness of the romance paradigm. The anomaly is not an uncommon one. It is familiar, no doubt, to those feminist critics like myself, conversant with deconstructive theories of romance yet still unable to hold back a tear when Mel Gibson marries his female co-star at the end of the movie. However, a closer analysis of Rich's autobiographical fictionalisations reveals that her rewriting is not shaped simply by the inability to reject an engrained cultural model. Rather, the discourses of romance play a more complex and self-justificatory role. In particular, by rewriting her life in terms of romance, Rich endeavours to explain (both to herself and possibly to an implied reader) how she was duped by the ideology of romance into marrying Charles and turning away from her family.

There is little doubt that the rift within the Boyle family which preceded Rich's marriage proved a source of constant regret well into her later years. Continual reference is made in the diary to the 'undutiful and disobedient' way in which she had forced her father to consent to the marriage (*Diary*, p. 123). I want to argue that Rich's romantically inflected autobiography can be understood as her personal project for *recovering* from this major upset. Romance, which carries its own particular brand of morality, is the discourse which most credibly authenticates Rich's decision to contract an unpopular marriage. It explains her mistakes, and her experience of familial estrangement, by highlighting the domination of a mythic system within which an individual has few claims to personal autonomy.

The justificatory techniques that characterise Rich's self-representations are easily apprehended in the attention given to the slow growth of her love. Following his first appearance as one of her brother's friends, Charles is only later viewed as a possible suitor. Rich utilises the conventional trope that love appeared when she least expected it. As she says of Charles:

> I began to find ... that my kindness for him grew and increased so much ... he did insensibly steal away my heart and got a greater possession of it than I knew he had. (*Autobiography*, p. 4)

The passage exteriorises the entanglement in such a way as to absolve Rich herself from romantic responsibility, and this is underscored in the deflective power of words such as 'possession' and 'steal'. And it is Rich as an unwitting victim, caught in circumstances beyond her control, upon which the rest of her narrative concentrates.

The fact that Rich is in a situation beyond her control works to mediate her part in the ensuing deception: she deceives her family and friends by embarking on a clandestine relationship composed of secret meetings and illicit liaisons, subterfuges necessitated by her father's opposition to the match. No doubt in practice Rich would have been all too aware of her father's feelings; in her narrative, however, his objections appear only *after* the couple has fallen in love, and hence are relegated to the status of a formulaic plot twist. Notably, it is as an obstacle to consummation that her father's disapproval is initially constructed. Charles's first declaration of love significantly coincides with the onset of Rich's anxieties over the *senex's* reactions:

> I did not find his declaration of his kindness disagreeable to me, but consideration ... made me sadly apprehend my father's displeasure. ... For my father, I knew, would never endure. (*Autobiography*, p. 4)

The perspective placed upon paternal opposition has the effect of persuading the reader to recognise romance as the dominant force propelling the narrative forward. From this vantage point, Rich's construction of her father's intervention as a formulaic device places confines upon his displeasure and encourages a reader to judge it adversely. The very familiarity of the predicament would seem to bode well for an eventually triumphant romantic resolution.

The possibility of an eventual resolution is lent additional support from the way in which the relationship between Rich and Charles develops via a series of chance happenings and fortuitous interventions. Beyond the care with which Rich elaborates the random nature of these encounters lies a romantic conception of destiny and fate: it is as if there is a hidden hand guiding (or indeed pushing) Rich toward the marital altar. Once again, the net result is to stress the primacy of sentimental emotion in the storyline and consequently to cast the behaviour of Rich's father in an aberrant light.

At a deeper level of the narrative, a stress upon the injustice of the father's reactions is taken up in a penetrating analysis of the importance of material goods and commodities. Paternal opposition to her match is constructed by Rich as an antipathy towards Charles's unimpressive economic credentials. (Although his father, the Earl of Warwick, was the largest landowner in Essex, Charles, as a younger son, was restricted to a paltry allowance).[15] Standing in stark contrast to her father's material worries is the 'purity' of Rich's feelings for Charles, which are encapsulated in reflections upon the suitor's impoverished status: 'when I married my husband, I had nothing of honour nor fortune in my thoughts; it was his person I married and cared for, not an estate' (*Autobiography*, p. 10). Implicit in such statements is a critique of her father's more financially inflected sensibility. The suggestion is that Rich, in her privileging of sublime emotions, enjoys the moral high ground.

The elevation of love over money is, of course, a crucial component of the romantic code of values. Interestingly, however, this is not represented as a code to which Rich always subscribes. Instead, at an early stage of the narrative, Charles's lack of money is cited by Rich as a reason for discouraging his advances: 'I considered my mind was too high, and I too expensively brought up, to bring myself to live contentedly with Mr. Rich's fortune' (*Autobiography*, p. 4). Only after a protracted process of internal discussion and debate does Rich move to a position from which she can prioritise love and minimise financial considerations. The disproportionate amount of time devoted to outlining both sides of the argument functions to highlight the difficulty of her decision and her horror at the prospect of alienating family and suitor alike. Dilemmas are given added impetus by the insertion into the narrative of passages of fraught vacillation.

In a sense, however, the battle is won even before it commences. The imperatives of the romance narrative (the primacy of sentimental emotion and the inevitability of heterosexual union) mean that familial attachments will be overruled eventually. Love figures as the key instrument with which Rich overcomes her fear of disobedience: 'the extraordinarily great kindness I had for Mr. Rich made me resolve to endure anything for his sake' (*Autobiography*, p. 9). Romantic love emerges simultaneously as an explanation for, and a legitimation of, filial estrangement.

Legitimating strategies are no less powerfully at work when Rich ultimately accepts Charles's proposal. It is entirely appropriate that the financial concerns that have haunted the narrative as a whole should reoccur at this climactic point, and in such a way as both to clarify and undergird the 'rightness' of her choice. Charles promises to 'make up to [Rich] the smallness of his fortune by the kindness he would have to [her]' (*Autobiography*, p. 5), and Rich declares that she would rather live with him in poverty and anonymity than with the richest man in the world: 'My kindness to him is such ... [I] ... should judge myself to be much more happy with his small [fortune], than with the greatest without him' (*Autobiography*, p. 6). 'Kindness', here a euphemism for 'love', renders unimportant the situation's material realities, and provides the concluding sanction for Rich's transgressive trajectory.

Throughout, her narrative has suggested the existence of a benevolent influence. Yet it is only as the courtship concludes that this influence is interpreted less as the inexorable working out of romance than as the force of God moving in his mysterious ways. Romance, in short, is constructed simultaneously as the guarantor of happiness and as the medium through which Rich is able to gain access to the deity:

> I by my marriage thought of nothing but having a person for whom I had a great passion ... yet [God] was pleased ... to bring me, *by my marriage* into a ... religious family. (*Autobiography*, p. 15; my emphasis)

The manoeuvre has a particularly paradoxical flavour. Even as she is conceptualising marriage as the perfect embodiment of romantic love, Rich suggests that her blessed union belongs to a grand scheme culminating in a relationship with another master. Romance becomes the necessary precondition for, and route towards, the religious. In this textual moment alone Rich's romantic rewriting of her life in the autobiography is held in delicate equipoise with the self constructed in the diary, a self which receives its shaping identity from contemporary discourses of exemplary biography.

*

Earlier I argued that Rich utilised the diary's potentialities with a view to securing comfort through private expression. Deploying the discourses of exemplary biography in the diary permits Rich to progress much further in this direction, to seize an opportunity to rewrite the experience of quotidian unhappiness in the guise of tragic martyrdom. Two examples of her self-flattering, strategic reinventions, one from 1667 and the other from 1673, attest to the ways in which Rich creates for herself a saintly identity:

> After dinner, my lord fell into an exceeding violent passion with me, and, in his passion, spoke many bitter and unkind words; I was in no fault to provoke him to it, but, I was enabled to bear his passion without saying anything to provoke him to continue it. (*Diary*, p. 154)

And

> After dinner, my lord sending me word ... that he was much offended with me, though my conscience cleared me from having done any thing to deserve his being so, yet I resolving to obey God who had bid me overcome evil with good, before Dr. W[alker] did tell him how troubled I was that he was displeased with me, and begged his pardon for what he thought I had done amiss, and promised to endeavour to avoid doing anything to offend him for the future.[16]

In the entries one can trace a trajectory leading from an initially resigned acceptance of domestic abuse to a more divinely sanctioned and actively engaged argument for a state of prevalent distress. The second quotation, in particular, plays up Rich's self-appointed role as a figure of patient forbearance. A Christ-like figure whose tribulations can be traced not to her own but to her husband's misdeeds, Rich constructs herself as the willing recipient of pain, as a singular example of the trials of virtue.

The ease with which Rich is able to construct a martyrology is related to the fact that the exemplary biography offers not only a validation but also a rationale for domestic friction and marital strife. Suffering is transfigured into a meaningful and necessary form of endurance, a valuable 'learning experience'. Rich's investment in this interpretation plays itself out in the ways in which frequent broodings on troubled circumstances are balanced by formulaic observations. For instance, after tolerating Charles's rages, she is prone to state: 'It was for my profit and good . . . I was enabled to justify God's proceedings with me, and to say, it was good for me that I was afflicted . . . And I was mightily encouraged' (*Diary*, p. 190). The rationalisation of her husband's controlling temperament facilitates a translation of earthly anger and despair into a higher spiritual plane. Similarly, a thankless night passed in nursing an abusive Charles is rewritten as a character-building exercise: 'by seasonable afflictions [God] had done me good', Rich comments, a realisation that 'did encourage and enable me patiently to bear his afflicting hand' (*Diary*, p. 242).

In the martyrological tradition, affliction is a prerequisite for salvation, and exemplary biographers were quick to point out that their heroines were assured of a cosy future. From this perspective, one can posit that, in contrast to romantic discourses which allow Rich to look back, exemplary discourses create the opportunity to glance ahead. For Rich, heaven is constructed both as a deliverance from daily troubles and as a reward for enduring them. Death is almost always represented as a pleasant prospect, if only because it entails the possibility of a spiritual destiny very different from that which her husband, she believes, is likely to experience. After an evening spent listening to the vulgar conversation of her husband and his brother, Rich consoles herself with the arguably uncharitable thought that:

> When I came to heaven I should never more have my soul grieved at the hearing of the filthy communication of the wicked, but instead of their loose and profane jests, hear the hallelujahs of the blessed . . . I considered that one moment in heaven would make me forget all the crosses of this life. (*Diary*, p. 191)

Seeing herself through the lens of another crucifixion, and consoling herself with the conviction that her tribulations are the passport to eternal happiness, Rich seems to find a healthy measure of comfort in the conviction that her husband is an unlikely candidate for heavenly glory.

At one and the same time, however, Rich utilises discourses of exemplary biography according to a set of more pragmatic imperatives. Modelling her existence on an exemplary life, in short, is also the mechanism whereby Rich justifies finding an alternative centre for her energies. This involves an implicit interrogation of the roles (wife and mother) that contemporary morality required her to serve.[17] A striking feature of her narrative is the way in which Rich almost always parallels her care for her husband with her forced neglect of godly duties.

Comments on the obligation to attend her husband's sickbed are frequently followed (usually in the same sentence) by references to having to miss divine worship:

> In the afternoon, my lord had a fit of an ague, and *so I was kept from being retired* by my attendance upon him . . . This day my lord was taken ill with the gout in great violence; I was constant in my attending upon him, *but had frequent returns to God by short ejaculations*. (*Diary*, pp. 77, 185; my emphasis)

The tension in the passage arises from the way in which Charles and God are situated in vexed and explicit opposition. Informing Rich's remarks is the perennial dilemma of the exemplary woman – the difficulties of reconciling a holy life with the family's expectations.[18] The extract imagines a conflict between wifely duties (attending to an ill husband) and godly duties (spending time in prayer). Nor is Charles the only one who is constructed as negatively competing for Rich's attention. In keeping with her representations of Charles's demands, Rich's evocation of her domestic responsibilities is likewise marked by a language of compulsion and burdensome necessity:

> I having heard of some disorder among some of my servants, I was *forced* to spend most of the morning in reproving and counselling them . . . In the afternoon, I was *hindered, by necessary employments*, from being retired. (*Diary*, pp. 199, 76; my emphasis)

Once again, the syntactical arrangement of the prose points to the other, more attractive location of personal communion, which domestic disorder threatens to displace. Because she describes husband and household in an identically idiomatic manner, she bestows upon both a similarly levelling gaze.

Given the preoccupation of many exemplary biographies with these and similar conflicts, such levelling could be dismissed as rhetorical were it not for the fact that a similarly antithetical arrangement is absent from Rich's descriptions of time spent with female friends and relatives. In these cases, instead of a syntactical opposition between two polarities, we find a sentence structure characterised by a free-flowing and accumulative narrative style. As Rich states in an entry dated 28 November 1666:

> After dinner, went to visit my lady Bedford; I had with her lord and her a great deal of good discourse. Afterwards went to see the duch-

ess of York's two children; had with the duchess a great deal of good discourse. Returned not home till late in the evening. After supper, committed my soul to God. (*Diary*, pp. 90–1)

Notable in this passage is the harmony between time spent with friends/family and communication enjoyed in divine company. On these occasions, a vital earthly and spiritual balance is secured. Such an agreement between conflicting demands is constructed by Rich as remote from her experience as a divine servant and a dutiful wife and house-keeper.

Extending the suggestion that God and Charles are generally opposed is the idea that they are locked in a specific rivalry for her love. On 11 April 1668 Rich notes: 'I had a sweet enjoyment of God this happy morning, for some two hours together whilst my lord slept' (*Diary*, p. 152). It is as if Rich participates in a clandestine union behind her husband's back. The notion is more forcefully elaborated in passages in which Rich mourns the extinction of romantic love:

> God is ... so merciful ... as to throw down our idol and his rival, and to make it ... seem withered in our eyes by having a less blinded passion for that person. (*Diary*, p. 262)

It is striking that Charles is depicted as an 'idol', since the designation implies that, whereas romantic love is a form of idolatry, devotion to the heavenly father is the true religion. God's is the fructifying, empowering love, in contrast to Charles's passion, which can only have a debilitating, enfeebling effect.

At times, it seems, the impression provided by the diary is that Charles was obscurely conscious of occupying a rivalrous position. Interestingly, Rich constructs Charles as forcing her to transfer her thoughts away from God and onto her husband's shortcomings and provocations. In an entry marked 19 January 1673, she writes:

> Before going to bed, receiving from my lord ... a very angry and pro-voking message ... I found my self so much by passion disordered that I could not bring my mind into a composed frame to pray before going to bed ... but went to rest without committing myself to God.[19]

Clearly evident here is Rich's endeavour to rewrite the source of her frustrations. Anger directed at Charles is transposed onto her own inability to pray. Through such stratagems, Rich gains a certain control over a situation which is beyond her power; rather than vex herself over her husband, she chooses to worry about the damage to her relationship with God, an area of consciousness over which she can wield a satisfying if imperfect degree of influence.

It is certainly the case that Rich represents God as more than a merely spiritual object of interest. Of course, the displacement of earthly love onto a heavenly object was strongly encouraged in contemporary devotional literature of the period, with many women describing their relationship to God in a highly stylised romantic vocabulary. However, it is possible to speculate that the displacement taps a deeper level of the imagination. An extraordinary ardour characterises many of those diary passages which dwell upon Rich's practice of religious exercises, leading to the suspicion that some sort of erotic transference between Charles and God seems to have taken place:

> I could steadfastly believe that I was my Beloved's and he was mine . . . It pleased God . . . in a more than ordinary manner to draw out the strength of my affections; and . . . to find my heart inflamed with love to him . . . with desire I did desire him, and found I esteemed him before the whole world. (*Diary*, pp. 184, 185)[20]

Here religious and romantic languages exist in a mutually constitutive accord, suggesting that spiritual longings and romantic imperatives find their meeting place in the personal relationship between a male divinity and a female believer. In this case, at least, both romance and exemplary biography are implicated in a symbiotic and self-perpetuating partnership. But the effect is momentary. More generally, the tensions and contradictions accompanying the balancing-act between romance and exemplary biography work to confirm that, for Rich, neither model of the self comes near to capturing the full complexity of life experience.

*

The relationship, in Rich's work, between the autobiographical act and a fictional model illuminates the interrelations and transactions through which seventeenth-century women defined themselves according to a culturally legitimised story. On the one hand, Rich's self-representations are salient demonstrations of the conformist power of a dominant social expectation. Both her texts reveal the extent to which women were shaped by contemporary ideological norms. From Rich's conjuration of seventeenth-century gendered images, her reimagining of life as formula and her embracing of stiff morality emerges a potent impression of the insidious force of seventeenth-century behavioural expectations and anticipations. On the other hand, Rich's writings and rewritings show the ways in which women were capable of appropriating existing ideological norms, reinventing them for their own advantage. Rich's vehement self-defence in the face of marital and familial pressure, her critical exposure of the myth of romantic love and her centralisation and elevation of her own individuality demonstrate that within apparently conservative modes and genres there were opportunities for exploring alternative subject positions and enabling possibilities.

Whatever possibilities other women in the period may have been able to exploit must inevitably remain a matter of conjecture.[21] Rich's privileged background, hardly typical of women in the seventeenth century, is the determining factor behind the appropriative exercises that she activates and, more generally, behind the literary references that are peppered through her work. Few women in seventeenth-century England would have had such an educated background, such book availability and such leisure to write. Nevertheless, this chapter's significance may not end with literate women. It is possible to speculate that, since myths similar to those perpetuated in romance and exemplary biography proliferate in oral as well as written forms, this chapter's conclusions about the deployment of fictional models in Rich's autobiographical texts may have more general implications. Versions of the same story circulate in different ways in culture (through hearsay and visual literacy, for instance) and are not always confined to the restrictions of the printed page.[22] In the absence of a more substantial body of evidence about women's reading in this period, Rich's example may yet prove essential to reconstituting the illiterate women's response to the romantic heroine of a ballad, to the exemplary woman of a funeral sermon or to the common activities indicted in a parliamentary pronouncement. At the very least, Rich's example demonstrates that the gendered reading process in early modern England was for more complicated, resistant and varied than has hitherto been realised.

Notes

1 Carolyn G. Heilbrun, *Writing a Woman's Life* (London, 1989) p. 109.
2 Mary Rich has also been the subject of several biographies. See Charlotte Fell *Mary Rich, Countess of Warwick, 1625–1678* (London, 1901), and Sara Heller Mendelson, *The Mental World of Stuart Women: Three Studies* (Brighton, 1987). For contemporary accounts of her life, see Anthony Walker, *Eureka, or, The Virtuous Woman Found* (London, 1678), and Thomas Fulwar, *A Funeral Elegy*

upon the much lamented death of... Mary, Lady Dowager, Countess of Warwick (London, 1678).

3 Mary Rich, *Memoir of Lady Warwick: Also her Diary* (London, 1847). All further references appear in the text. At those points where this edition of the diary is incomplete, I cite from the complete diary version, which is still in manuscript.

4 Mary Rich, *Autobiography of Mary, Countess of Warwick*, ed. T. Croker (London, 1848). All further references appear in the text.

5 For a discussion of the simultaneously public and private nature of seventeenth-century women's autobiography, see Helen Wilcox, 'Private Writing and Public Function: Autobiographical Texts by Renaissance Englishwomen' in *Gloriana's Face: Women, Public and Private, in the English Renaissance*, ed. S.P. Cerasano and Marion Wynne-Davies (Hemel Hempstead, 1992) pp. 47–62.

6 This is suggested by a diary entry which claims that, on the 8, 9 and 10 February 1671, Rich was recording, in her words, 'the specialities of [her] life' (*Diary*, p. 231).

7 Other commentators have noted the disparity between the two texts. See *A Biographical Dictionary of English Women Writers, 1580–1720*, ed. Maureen Bell, George Parfitt and Simon Shepherd (Hemel Hempstead, 1990) p. 165.

8 *Revelations: Diaries of Women*, ed. Mary Jane Moffat and Charlotte Painter (New York, 1974) p. 5.

9 Charles's violent temperament is attested to in other sources. See Walker, *Eureka*, p. 4. For a discussion of domestic violence in the early modern period, see Susan Dwyer Amussen, '"Being Stirred to much Unquietness": Violence and Domestic Violence in Early Modern England', *Journal of Women's History*, 6 (1994) 70–89.

10 See Jacqueline Pearson, 'Women Reading, Reading Women', in *Women and Literature in Britain 1500–1700*, ed. Helen Wilcox (Cambridge, 1996) pp. 80–99.

11 British Library, Boyle MSS, XVII, no. 75. This is also referred to in Mendelson, *Mental*, p. 66, and Smith, *Mary Rich*, p. 47.

12 These references seem to suggest that Rich was a consumer of the French heroic romances so popular in England at this time. Examples include Jean Desmarets de St. Sorlin, *Ariana* (London, 1636); François le Metel de Boisrobert, *The Indian History of Anaxander and Orazia* (London, 1639), and Gilbert sieur de Verdier Saulnier, *The Love and Arms of the Greek Princes or the romant of romants* (London, 1640).

13 Examples include Samuel Ainsworth, *A sermon preached at the funeral of that religious gentlewoman Mrs Dorothy Hanbury* (London, 1645); Thomas Heywood, *The exemplary lives and memorable acts of nine of the most worthy women of the world* (London, 1640) and John Ley, *A pattern of piety* (London, 1640).

14 British Library, Additional MS. 27353, fo. 36v.

15 For the predicament of the younger son in early modern England , see Linda Pollock, 'Younger Sons in Tudor and Stuart England', *History Today*, 39, June (1989) 23–9.

16 British Library, Additional MS. 27353, fos. 18v, 31v.

17 On this subject, see Patricia Crawford, *Women and Religion in England, 1500–1720* (London and New York, 1993) p. 90.

18 These duties were set out in such books as Robert Cleaver, *A godly form of householde governement* (London, 1598) pp. 218–35, and William Gouge, *Of domesticall duties* (London, 1622) pp. 267–348.

19 British Library, Additional MS. 27353, fo. 27^r.
20 Mendelson wittily notes that Rich herself may have 'recognised an earthly tincture to her more rapturous expressions, since she crossed them out at a later date' (*Mental*, p. 107).
21 On the dangers of generalising from the example of privileged women, see Margaret Ferguson, 'Moderation and its Discontents: Recent Work on Renaissance Women', *Feminist Studies*, 20 (1994) 349–65.
22 For recent work on the circulation of oral narratives, see Elizabeth Tonkin, *Narrating Our Pasts: The Social Construction of Oral History* (Cambridge, 1992), and Paul Thompson, *The Voice of the Past: Oral History* (Oxford, 1988). See also the discussion of autobiography and ballad in Henk Dragstra's essay in this volume.

11
Last Farewell to the World: Semi-oral Autobiography in Seventeenth-century Broadside Ballads

Henk Dragstra

Autobiography and Orality

Seventeenth-century England saw autobiography spreading out, or trickling down, from privileged individuals to the common citizen. This development mirrored the spread of literacy, and followed in its wake, but not instantly: there can be no autobiographical writing without 'a sense of one's own importance as an individual'.[1] Most study of seventeenth-century autobiography has concentrated on autobiographies as products of literacy, which implies that the nascent sense of autobiographical self that is encountered has crossed the threshold between the literate mindset and the oral. But what about all those masses whose literacy, whatever it may have been, stopped short of literary self-expression? Was there no such thing as a sub-literate, or semi-literate, or semi-oral sense of autobiographical self?

Many individuals took part in literate communication mainly as listeners to texts that were sung, read aloud or enacted: the unlearned who went to church but read little Scripture, the 'groundlings' in the theatre, the *mobile vulgus* that read no news bulletins but was keen not to miss a pageant or public execution. Historical scholars have studied the 'last dying speeches' made by convicts on the scaffold: such speeches, public, confessional and made in a dramatic setting, must have been the most powerful example of non-secular autobiography that a metropolitan crowd would hear regularly.[2] Much encouraged by the authorities provided they sounded sufficiently contrite, these effusions, no doubt heavily edited, were disseminated all over the kingdom in various

166

forms of cheap print. Prose accounts, printed in pamphlet or chapbook form, travelled far and wide;[3] ballad versions went even further, reaching the illiterate as well as the learned.[4] Hawked in cities and villages, pasted over the chimney in taverns and cottages, such broadside ballads could be sung from print as well as memorised orally. Here are some characteristic stanzas, supposed to be uttered by a printer and a midwife respectively:

> I was a Gentleman by my de[s]cent,
> But to Extravagancy always bent:
> And my Estate I quickly did consume,
> But did not think of this my Fatal Doom.
> (II, 154)[5]

> For three and thirty years ago,
> I Midwife did begin.
> And of late years astutely know,
> I have been murdering.
> Sweet Infants from their Mothers Womb,
> Oh! Wretched Creature, I
> Starving did make their Dismal Doom,
> For which I now must dye.
> (II, 192)

I shall refer to these songs as 'autobiographical execution ballads'. The most salient elements that they have in common with autobiography as we generally know it can be summarised as literacy, historicity, biographicality and autodiegesis.

That these ballads are products of literacy is undeniable: they are printed and they have titles. A representative case is the ballad of the printer quoted above, which I here reproduce in typography as close to the original as possible:

DISNY'S Last Farewell.
Being an Account of the Execution of VVilliam Disney Esqr, **who was Drawn, Hang'd, and Quartered, On** Monday **the 29th of this Instant** June, 1685. **For Printing of Monmouth's Treasonable** Declaration
(II, 154)

The publishers' names appear at the bottom of the same sheet: 'Printed for J. Clarke, UU. Thackeray, and T. Passenger', and the broadside also features formulas of imprimit and registration: '*This may be Printed,*

R.L.S. July *the* 1st, 1685', and 'Entered according to Order.' Crude wood-cuts depicting gruesome executions help to draw the buyers' attention. However, the text itself reminds us that we are not dealing with a purely literate text, but with a song: 'To the Tune of, Fortune my Foe.'

As the above example shows, these songs do not concern fictional characters but historical individuals whose names and verifiable particulars are given, as well as details of the place and/or moment of their execution. Most of the individuals mentioned in such ballads can be traced from other historical sources, and this goes for commoners as well as the aristocracy. Documents confirming their historicity include legal records, pamphlets in prose and political sources.[6]

These execution ballads are biographical: they review individual lives. This may be done in detail, as in 'Sir Walter Rauleigh his lamentation: Who was beheaded in the old Pallace at Westminster the 23 of October. 1618' (I, 110), which is 23 six-verse stanzas long; or it may be quite brief, like 'Disny's Last Farewell', which consists of 16 four-verse stanzas. From the scaffold perspective, the lives that these songs review are very much foreshortened: childhood and puberty tend to go unmentioned, crimes figure large, and everything is steeped in self-accusation and repentance.

This constant emphasis on guilt is very much reinforced by the oral aspects of the songs. Since they are set to easy and popular tunes, they are in stanzas, and a favourite way for stanzas to end is in a 'burden', a recurring phrase, like a refrain. Typical burdens of execution ballads are: 'for all my Villany' (II, 161), and 'A sad Example made' (II, 194); burdens ending in the word 'die' are especially common. This fixed ending to each stanza not only means a constraint on the rhyming, but also on the content of each stanza and, therefore, on the autobiographical structure. Every description of an event, every thought on the subject's part, must, within the space of eight verses, lead to the same condemnatory moral.

The ballads are also autodiegetic,[7] that is, the narrative is in the first-person singular. The relation between narrator and agent is basically one of repentance: the narrator bewails his or her crimes, and points to the impending or present execution. Disney's ballad, for instance, opens with 'You Loyal Subjects wheresoe're you be, / See here the Fruits of my Disloyalty; / And let my shameful end a caution give / To all, that they in Loyalty may live.' This moral emphasis tends to be reflected in the printed captions of the broadsides: often the individual's name appeared small, while the moral role he was made to play featured large. This difference in scale is very striking in, for instance, 'The last dying Lamentation of Thomas Randall' (II, 161); that line is dwarfed by the overhanging headlines 'THE / Mournful Murtherer'. The genre is generally serious, indeed draconic, in tone, with its heavy emphasis on repentance in the face of death, and the sudden awareness that 'A viler sinner lived not then I' (II, 147).

Claims of being 'the chief of sinners' are familiar enough from religious autobiographies of the time, as well as from heterodiegetic reports.[8] Several eyewitness accounts of public executions of robbers mention that the offender addressed the crowds with edifying remarks.[9] Yet to a modern reader the identity of narrator and agent in these ballads is by no means a guarantee of authenticity; indeed, from a modern point of view, the bulk of these texts may be termed 'pseudo-autobiography'. When an execution was to be expected, publishers saw their chance to make a profit; they would hire a hack to write words to a popular tune, and had their product hawked where and when the execution took place. Claims of genuine authorship are soon suspected and dismissed by the modern reader, even without consulting historical evidence, because we tend to find the texts unconvincing. They contain far too many incongruencies for us to take them at face value.

Grotesque effects are inherent in their close link to the place and time of the execution described. In several cases, as in Raleigh's, the subject is actually represented as speaking at the very moment of dying: 'My head on block is laid, / And my last part is plaid: / Fortune hath me betraid, / sweet Jesus grant mercy' (stanza 22). An 'vnnaturall Wife', the murderess of her husband (I, 122) says, in the past tense, 'And there in Smith-field at a Stake / my latest breath there did I take.' We are evidently dealing with a literary convention, offering the poet maximum scope for dramatisation and popularised by speeches of historical ghosts like those in *The Mirror for Magistrates*. As Rollins observes, 'verisimilitude was the least of the ballad-writer's troubles'.[10] This fact did not escape contemporary notice: James Shirley satirised it in his play *The Court Secret* of 1653:

> MEN: What's that?
> PEDRO: A ballad sir,
> Before I die, to let the people know
> How I behav'd myself upon the scaffold;
> With other passages, that will delight
> The people, when I take my leave of the world,
> Made to a pavin tune. – Will you hear it?[11]

Moreover, these ballads often feature striking gaps in point of view between the narrator and the agent of the crime, for instance in 'The

Lamentation of Master Pages wife of Plimmouth' (I, 126). According to the broadside caption, the young woman, 'being enforced by her Parents to wed him against her will, did most wickedly consent to his murther, for the loue of George Strangwidge: for which fact she suffered death at Barstable in deuonshire.' Though its message is a plea against forced marriages, the poet shows little empathy. Sold into marriage to a man vastly her senior, the woman is made to voice her aversion to him thus: 'Cause knew I none I should despise him so, / That such disdaine within my minde did grow, / Saue only this that fancy did me moue, / And told me still George Strangwidge was my loue.' Claimed to have been 'written with her owne hand, a little before her death', the text sports a very rhetorical style, dripping with alliterations and parallelisms; so does her 'Sorrowful Complaint' (I, 127), which should not surprise us; but 'The Lamentation of George Strangwidge', her accomplice, printed on the same sheet, is in the same distinctive style. Particularly improbable is the sub-genre's pervasive craving for harsh punishment of self, in phrases like 'The Laws cannot be too severe / for such a Wretch as I' (II, 192).

This modern impression of incongruity is supported in some cases by historical evidence. 'The Counterfeit Coyner: or, The Dying Lamentation of John Moor, the Tripe man, who was Arrained and found Guilty of Counterfiting the Coyn of the Kingdom, and was accordingly Executed at Tyburn, on Fryday the 12 of July, 1695' (V, 7), is in the familiar form of a confession and repentance; but a contemporary witness, convinced of his guilt, reports how Moor contended his innocence throughout his trial, imprisonment and execution.[12] Thomas Randall, a highwayman, acknowledged he was guilty of many robberies, but denied to the last the killing of one Roger Levins;[13] nevertheless, in a ballad about his execution (II, 161), he is made to confess this murder in specific detail.

Particularly unconvincing, and demonstrably false, are songs about the imaginary 'Popish Plot' of 1678 and 'Rye House Plot' of 1683, in which the executions of alleged conspirators are exploited for propagandistic purposes.[14] In a ballad mostly told in the third person, Edward Coleman, a Jesuit convicted on trumped-up charges of conspiring in the 'Popish plot', is made to say, as he is about to meet his Maker: 'Nor let the proud Prelate of Rome, nor his train, / Choose Engines of mischief, whose Warrants are vain'.[15] When Lord Russell was executed on the accusation of high treason for taking part in the 'Rye House Plot', he made a last speech and also delivered into the hands of the Sheriffs a written declaration (both naturally in prose). The speech, in which he asserts his innocence,[16] is partly quoted in a satirical ballad (*RB* V, 324);

but this did not forestall the production of another, completely fictitious ballad (*RB* V, 691), in which he makes a full confession. But even when the political acts that such ballads enumerate are genuine, the ballads made about them are evidently 'ought-to-biography': they do not express what the subject felt about them, but what the public, or the authorities, felt they should feel.

The falseness of the autodiegetic form of these ballads is often obvious in the sloppiness with which it is observed. In many cases, we find the autodiegesis interrupted or taken over by another voice, a commentator who is the poetaster of the piece, or a moralistic persona. Such authorial interference is a common feature not only of execution ballads, but of broadside ballads generally. But even while ostensibly preserving the autodiegetic illusion, the rhymester's hand often shows in narrative inconsistency. The highwayman Francis Winter (II, 188) says about an alleged murder he committed, 'Whether I kill'd the man or no, I cannot fully say'; evidently, the ignorance expressed here is the poet's, not the highwayman's. And the 'Bloody Miller' Francis Cooper says about himself: 'I was a Miller by my Trade, / it plainly doth appear' (II, 156). But if such incongruencies disturb us, the hack poets did not care and neither did their audience.

The elements summed up above combine to make this kind of ballad autobiographical enough in form to represent to its consumers an image of individuals reviewing their own life. To most of the semi-literate public, ballads were the only readily accessible printed source of news, and few were likely critically to compare two conflicting printed interpretations of one execution. Though ballads might be contradicted in the streets, squares and taverns where they circulated, comments were as ephemeral as any spoken word. Actual statements given out by the executee were easily forgotten or drowned out by fictional and satirical representations, in ballad form or otherwise. To its intended audience, the distinction between an autobiographical and a pseudo-autobiographical execution ballad was not an issue. The spectacle of the subject's execution probably did more for a ballad's credibility than poor or slanted writing could spoil.

As Natascha Würzbach has noted, 'role-playing', the concept of impersonation, is essential to many 'street ballads'.[17] Historical characters long deceased, such as Jane Shore (I, 486–7), fairy creatures like Robin Goodfellow (I, 80), the Cheapside Cross (I, 66–7), and even a man's back (I, 446), were pressed into service by the balladeer as mouthpieces for some political or moral message. In practice, this mouthpiece function fell to the balladmonger, and at a later stage to the customer.

The public was used to hearing ballad-singers impersonate unlikely speakers without change of dress, though probably with change of voice or demeanour, however poorly done. Nothing was more natural than that the singer should slip in and out of impersonation in his or her singing.

This casual attitude displayed by poetasters, singers and audience towards the narrative framework of execution ballads corresponds to a well-known style feature of oral literature, and especially of folk ballads: its tendency to spice narrative with dialogue without the aid of 'he said' and 'she answered'. The oral memory deals in actions and structures, oral narrative proceeds by leaps and bounds; but this does not spoil the audience's enjoyment, for in a good performance the identity of the speaking subject is evident from the performer's tone of voice, expression and gestures. Obviously, the same applies to the singing of ballads from broadsheets.

Printed text though the ballad may be to us, its performance was oral; and so was its style of characterisation. In ballads that show evidence of oral transmission, narration is starkly hyperbolic and dramatic, characters are drawn in bold strokes, and their actions speak for them more than their words. And although 'street ballads' express feeling much more verbosely than standard oral ones, they reveal a similar attitude towards character. If their lyrics represented public indignation rather than the individual conscience, that made them all the more popular. In the world of semi-literacy, self-revelatory monologues by stage villains were as familiar as ranting Herod in medieval plays; the London street public eagerly absorbed the stereotyped roles into which ballads cast notorious individuals. The broadside ballad catered to crude concepts of character, and was, as Claude M. Simpson put it, 'both a conditioner and a mirror of popular attitudes'.[18] In the seventeenth century, it takes us as close to the semi-literate secular autobiographical image of the self as we can hope to get.

Autobiographical Execution Ballads in Their Time

Execution ballads were surprisingly rare in the early seventeenth century; after the Restoration they multiplied, and a perusal of the most accessible editions will easily yield a hundred instances of the later type. The most usual scholarly term for them is 'Goodnight' or 'Last Goodnight'.[19] But whereas scholarly usage tends to favour fixed terms, the seventeenth-century broadside ballad was no more static than was contemporary society. Developments are reflected in the generic terms by which the ballad texts refer to themselves, especially by those used in the headlines, and these in turn affect the projected image of the autobiographical self.

The term 'Goodnight' is associated with one of the best-known English broadsides, 'A lamentable new ballad vpon the Earle of Essex his death' (I, 106), set to a tune called 'The Kings last good night'. The song, which heaps praise on the popular Earl, is in eight-verse stanzas, each ending in the burden 'last goodnight', or variations of it. It consists mostly of third-person narrative, but it includes a scaffold apologia by the Earl 'himself'. In another imprint (II, 163), the tune has assumed the name of 'Essex' last Good Night.' The same name is applied to the tune of 'A Sorrowfull Song, Made vpon the murther and vntimely death of Sir Thomas Ouerbury Knight' (V App, 9, 10). Both songs belong to the early seventeenth century: Essex was beheaded in 1601, and rumours about the poisoning of Overbury became public in 1615. William Chappell claimed that the ballads sung to *Essex's Goodnight*, or to its companion tune *Welladay*, are 'very numerous';[20] this is without foundation in fact for the *Goodnight* tune. The further instances Chappell mentions for it number only three; Claude M. Simpson mentions three more.[21] All of these ballads using the tune, except a Scottish one, are of the early seventeenth century; none is remotely like an autobiographical execution ballad;[22] rather, they are associated with disasters and with royalty.

The word 'Goodnight' is hardly ever used in broadside titles. 'John Armstrong's Last Good-Night. [d]eclaring how Iohn Armstrong and his eightscore men fought a bloody bout with the Scottish King at Edinborough' (II, 133) contains both heterodiegetic and autodiegetic narration; its hero's last words are martial rather than autobiographic. Armstrong met his death in 1528 or 1529, and the ballad, probably composed soon afterwards, shows signs of oral transmission; indeed, Francis James Child included it in his famous collection (as number 169).[23]

Thus in the early decades of the seventeenth century, and more probably before that,[24] the term 'Goodnight' may have been used for a few broadside ballads; but its use petered out in the course of the century, as far as this genre is concerned. In oral tradition it was more prominent. Besides the ballad of John Armstrong, mentioned above, there are 'Lord Maxwell's Last Goodnight' (Child 195) and 'Lord Derwentwater's Goodnight' (Child 208). What the literate and oral examples have in common is not public execution but protagonists of tragic illustriousness or bravery;[25] and this may explain why the term's use dwindled as broadside execution songs proliferated. With its noble overtones, the term must have been considered unsuitable for the last words of common criminals or traitors. Thus the term as used by authoritative scholars such as Chappell and Rollins may reflect usage of earlier or later centuries,[26] but it certainly does not correspond to notions of genre current in the course of the seventeenth century, which in this case seem to have been notions of class and of decorousness.

In the captions of the ballads as printed, the term most frequently used for autobiographic execution ballads is 'Lamentation', as in 'Sir Walter Rauleigh his lamentation' (I, 110), which has been mentioned before. Other terms used are 'Complaint' and 'Confession', as in 'The sorrowful complaint of Mistris Page' (I, 127), mentioned above; and 'Capt. Whitney's Confession: or, his Penitent Lamentation, Under a sence of a Guilty Conscience, on the Day of his execution at the Porter's Block, near Smithfield-Bars, which was on the First of February, 1693' (II, 186). There is also the term 'Last Farewell', as in 'Disny's Last Farewell' (II, 154), which has been mentioned above.

Terms like 'Lamentation' and 'Complaint' have a literate pedigree reaching back to the sixteenth century, or even to the Middle Ages;[27] and we are not surprised to find them applied to a wide variety of songs concerning generalised or fictional speakers, both grave, such as 'The sorrowful Lamentation of a penitent Sinner' (II, 13) and bawdy, for instance 'The Maids Complaint For want of a Dil doul' (IV, 50). The two terms are far too imprecise to serve a generic purpose. 'Confession', although familiar to us from St Augustine's autobiographical work, suggests more specific links with public executions, or rather, their judicial foreplay. It is used only rarely, but the instances that do occur, such as 'The Midwife of Poplar's Sorrowful Confession and Lamentation in Newgate' (II, 192), are very good examples of the autobiographical execution ballad. The phrase 'Farewell', straighforward and colloquial as it may seem, equally has many literary connotations: it features in the titles of many songs of frivolous tone, and often in a neoclassical pastoral vein. Like 'Lamentation' and 'Complaint', it can refer to many occasions, amongst which death is only one.

A much surer indication of finality is, naturally, provided by the word 'Last'. Coupled with this adjective, both 'Lamentation' and 'Farewell' are situated firmly in the forecourts of death. The collocation of 'Last' and 'Lamentation' occurs, for instance, in 'The Mournful Murtherer: or, The last Dying Lamentation of Thomas Randall' (II, 161); whereas 'Farewell' occurs almost always as 'Last Farewell' when it is the title of a sung speech at death's door, as in 'Disny's Last Farewell' (II, 154). Where such combinations refer to executions, they are remarkable for their homogeneity and autobiographical aspects of the ballads, as well as for the 'democratic' way in which they are used: high-born 'plotters' such as the Duke of Monmouth (II, 243) and Sir John Fenwick (V, 22) are set on a par with a bunch of pirates of Captain Every's crew (II, 199) and with the highwaymen William Davis (II, 187) and Francis Winter (II, 188).

With the demise of the 'Good-night' broadside ballad in the course of the seventeenth century, the pattern of emotional response aimed at in Last Lamentations and Farewells affords us an interesting glance at developments in public attitudes towards individual suffering. Instead of the tragically heroic mode associated with a popular but fallible aristocrat like the Earl of Essex, later 'autobiographic' execution ballads present their subjects as out-and-out criminals. The association of this mode with the term 'Last Farewell' is in marked contrast to farewells in oral tradition such as 'The Death of Parcy Reed' (Child 193); 'Jamie Douglas' (Child 204), and 'Flodden Field' (Child 168), which are as heroic in tone and setting as the oral 'Goodnights'. This contrast between the censorship-free oral ballad and the broadsides suggests that in the later seventeenth century, censorship had no tolerance for the depiction of heroic and tragic rebels like Essex and Raleigh.[28] To rebel was to be a 'plotter', that is, a traitor, and traitors deserved only humiliation.[29] This principle was applied with revolting rigour as block and axe yielded to the obscenity of the hanging, drawing and quartering ritual. The hurdle ensured that 'plotters' arrived at the place of execution looking suitably dishevelled, dusty and dazed; the gruesome spectacle and overwhelming stench of the exhibited quarters served to erase all sense of dignity afterwards.

Terms used in titles of execution ballads conform sufficiently to a pattern for us to recognise a nascent awareness of autobiography as a distinct mode. The phrases 'Last Lamentation' or 'Last Dying Lamentation', and 'Last Farewell', or 'Last Farewell to the World', are fairly reliable indicators of an autobiographic execution ballad. Execution-related Last Lamentations and Last Farewells together form the nucleus of one category, distinctive enough in its time to constitute a single sub-genre. What is more, they form a body of material exemplifying the seventeenth-century social expansion of the sense of an autobiographical self.

The Meaning of Tune

If we wish to understand what concepts of genre were in operation in seventeenth-century street-ballads, oral aspects provide important clues. Genre concepts in this semi-literate sphere are based on association rather than definition, and it is especially the associations between text, occasion and tune that should interest us. The more often a specific tune was used for a specific kind of text and occasion, the more prominent that tune would become as an attracter of attention, and therefore, of buyers. Thus certain tunes became favourite vehicles for certain kinds of text. Ballads about women were often set to the tune *The Ladies Fall*; songs in praise of the king to *Let Caesar live long*. This association between tune and occasion or subject-matter resulted in some tunes being considered 'filthy' or 'villanous' by individuals of taste. Admittedly, serious authors tended to consider all aspects of ballad-singing disgusting;[30] but there is good evidence that specific tunes were considered gross as such. For instance, in Shakespeare's *King Henry the Fourth, Part I*, Falstaff threatens to 'have ballads made on you all, / and sung to filthy tunes' (II, 1); Graeculo, in Massinger's *The Bondman* (1623), hopes he will be spared the fate of being executed twice, 'At the Gallowes first, and after in a ballad / Sung to some villanous tune' (V. iii, 245–6).

According to Hyder E. Rollins, the tune *Welladay* was 'the favourite tune for good-nights after it had been used in a ballad (1603) attributed to Robert, Earl of Essex'.[31] This refers to 'A lamentable Ditty composed vpon the death of Robert [Devereux] late Earle of Essex' (I, 107), a companion-piece to the 'Good-night' mentioned above. Like that ballad, it praises Essex highly; though basically third-person narration, it contains his 'autobiographical' speech. Another instance of the tune's use (1618) is 'Sir Walter Rauleigh his lamentation' (I, 110), which is truly autobiographical in style throughout, evoking pity rather than censure. Set to the same tune are 'The true manner of the life and Death of Sir Thomas Wentworth...beheaded the 12...of May, 1641', by Laurence Price;[32] 'King Charles His Speech, and last Farewell to the World...Ianuary 30. 1648,' which contains very few of the King's words and much praise of him;[33] and 'No Naturall Mother, but a Monster', about an infanticidal mother, licensed 1634.[34] What associations with executions – mostly beheadings – the tune might have, wore off in the course of the century. Two earlier instances of its use (I, 108–9 and I, 148–9), printed *circa* 1619 and 1631 respectively, feature neither public execution nor farewell speech. Four mid-century songs to this tune are cheerful rather than macabre.[35]

There is good evidence that the age knew the concept of the 'hanging tune'; the phrase is used by Rowley in his *Noble Soldier*, of 1634:

> The King! Shall I be bitter 'gainst the King?
> I shall have scurvy ballads made of me,
> Sung to the hanging tune!

William Chappell claims that this refers to *Fortune my Foe;*[36] Rollins takes it upon trust from Chappell.[37] Whether the inference is correct or not,[38] the tune certainly had moralistic overtones. 'The penitent Traytor: the humble petition of a Devonshire gentleman, who was condemned for treason, and executed of the same, anno 1641', refers to it as follows:

> How could I bless thee, couldst thou take away
> My life and infamy both in one day?
> But this in ballads will survive I know,
> Sung to that preaching tune, Fortune my Foe.

This quotation and the one from Rowley together reflect two aspects that we often find together in execution ballads: references to actual public executions, and moral admonitions to the general public.

The earliest preserved instance of the tune, already popular in the late sixteenth century, is entitled 'A sweet Sonnett, wherein the Lover exclaimeth against Fortune for the loss of his Ladies favour', and opens with 'Fortune, my Foe, why dost thou frown on me?' (I, 512). The tune recurs in 'An excellent song, wherein you shall find, / Great consolation for a troubled mind' (II, 63, registered 1656). But in the early seventeenth century, the tune may indeed be called a 'hanging tune': among its earlier instances, execution ballads figure large. They include the 'autobiographic' effusions of our friends Mr. Page's Wife (I, 126–7) and George Strangwidge (I, 127), executed in 1609; and of others[39] dating from 1615 (I, 130–1); from 1616 (I, 124–5) and from 1624 (II, 196).[40] The tune 'peaked' in the 1630s[41] with three execution ballads in autobiographical style and two with third-person narration.[42] After this its use for execution ballads petered out: 'Disny's Last Farewell' of 1685 (II, 154) was a lone straggler. It was, however, used for various other grim themes: the late-sixteenth century complaint of Titus Andronicus, (I, 86); the tragedy of Dr Faustus (II, 142), licensed 1624, but probably much older; the burning of Cork in 1622 (I, 68);[43] a 'judgement of God' in 1660;[44] and another fire in 1662.[45]

The tune changed its name, and perhaps its musical features,[46] during the seventeenth century: the 'excellent song' of 1656 mentioned above (II, 63) is set to *Fortune my Foe*, but opens with 'Aim not too High in things above thy reach'. As *Aim Not Too High*, the tune continued its life in nearly 30 ballads; but only in four cases can we speak of execution ballads. One of these describes in third-person narrative the beheading of Charles I (II, 204); two others concern criminals hanged in 1684 and

1685 (II, 153; 169); there is only one in the autobiographical style, about a hanging in 1691 (II, 200). In the other ballads, murder and deathbeds figure significantly; mortal wounds, a fatal hailstorm and poisonous snakes are also in evidence. But in these rhymed bulletins, morals not executions are the chief concern. So instead of hangings and other executions, the tune *Aim not too high*, as its name suggests, became associated in the course of the seventeenth century with moralism, evident in the very titles of some ballads, which feature terms like 'Caveat' (II, 36) and 'Looking-Glass' (II, 47, 68). Well may the ballad-rhymester call it 'that preaching tune', but 'hanging tune' was to become a misnomer: I have found only four post-Restoration instances of this tune being used for execution ballads.[47]

One factor probably contributing to, or influenced by, this tune's shift in significance is the usurpation of its function by a new tune which came into vogue in the final decades of the century: *Russell's Farewell*. In the year 1683 Lord Russell was executed for high treason, and as we have seen, this event was eagerly seized upon by ballad-writers. One of the songs written for the occasion was 'The Lord Russels Last Farewel to the World', beginning 'Farewel, farewel to Mortal Powers' (*RB* V, 691). This song came with a new tune, which was subsequently used for some 43 ballads. It makes a nervous impression, 'pulverising' the text's syllables into sequences of brief notes, unlike *Fortune my Foe*, which is much more stately and solemn.[48] This new tune soon became associated, in the public mind, with this particular kind of event, and this particular literary genre: 22 out of 43 ballads to the tune of *Russell's Farewell* are autobiographical execution ballads.[49] That is about half of the total, which makes the tune quite as effective an index to this semi-oral sub-genre as are printed headings, sometimes more so. This is especially striking in the broadside entitled 'King Jameses Royal Victory' (II, 237), to the tune of *Russell's Farewell*: as the tune hints, the text is in fact the 'treasonable' Duke of Monmouth's Last Farewell, followed by three final stanzas of authorial comment. Like the concept of 'Last Farewell to the World' or 'Last Dying Lamentation' in broadside titles, the tune was 'democratic': it would do for the Duke of Monmouth as well as for a couple of pirates. This egalitarianism would have pleased the authorities: from their point of view, robbers and rebels were equally obnoxious and detestable, and the association of the former with the latter by means of a tune provided another form of public degradation.

We have seen how other dignified execution tunes, *Essex's Goodnight* and *Welladay*, were well out of use by the middle of the seventeenth century. *Fortune my Foe* was 'split'; associated both with executions, moral lessons and tragic individualism, it was replaced by two tunes: *Aim not too high* as a vehicle for moral lessons to all citizens, rarely related to executions; and *Russell's Farewell* as a means of marking specific individuals as objects of hate and contempt. Indeed, *Russell's Farewell* must have been, to the public ear, both a 'villanous' and a 'hanging tune': when it was floating on the London air, there was an execution afoot. This was partly due to the burden, of course; I have found five instances with a burden on the theme 'now I'm Condemn'd to dye', and examples of burdens I have quoted above are similarly draconian. But even where there was no regular burden, the tune itself spoke a message. The tune would be a signal, like the sound of an ambulance or police siren in our day.

In the seventeenth century the autobiographical act, or what passed for it, was to the semi- or sub-literate London crowds an integral element of the most horrifying and humiliating torture imaginable. Recently, Michael Mascuch has argued that public executions and the printed merchandise that accompanied them in the eighteenth century tended to romanticise and glamorise the 'protagonist' of the event.[50] As we have seen, admiration of the real-life condemned is scarce in printed ballads of the seventeenth century, manifesting itself only in the early decades, and only in the case of aristocrats.

The tones of *Russell's Farewell* promised entertainment to the public; what they meant to the offender of the day we shudder to imagine; but to both parties, the tune-cum-text were autobiography, public humiliation and death knell combined. As Lacqueur shows, the authorities did not have complete control of everything that happened at actual executions; censorship, however, ensured that printed ballads gave approved versions of them, with the victim duly degraded and stereotyped.[51] The fate of being 'balleted' was dreaded at any event; it was to be made a public example and a public shame.[52] Associated with executions, it also made the object a public victim. Autobiography always involves a figurative sticking out of the neck; in the streets of seventeenth-century London this was literally true. To the semi-literate masses, composing your autobiography would have been like writing your own death warrant. As John Taylor the Water Poet put it:

> If any one (as I dare boldly done)
> His Birth, his breeding, and his Life declare:
> Let him appeare, and I dare lay my necke,
> He will be hang'd, or else deserve a checke. [53]

Notes

1   Paul Delany, *British Autobiography in the Seventeenth Century* (London, 1969) p. 108.

2   J.A. Sharpe, ' "Last Dying Speeches": Religion, Ideology and Public Execution in Seventeenth-Century England', *Past & Present* 107 (1985) 144–67; Thomas W. Lacqueur, 'Crowds, Carnival and the State in English Executions, 1604–1868', in *The First Modern Society*, ed. A.L. Beier et al. (Cambridge 1989) 305–55; Peter Lake, 'Deeds against Nature: Cheap Print, Protestantism and Murder in Early Seventeenth-Century England', in *Culture and Politics in Early Stuart England*, ed. Kevin Sharpe and Peter Lake (London, 1994) 257–83.

3   See Margaret Spufford, *Small Books and Pleasant Histories* (Cambridge, 1981); and Tessa Watt, *Cheap Print and Popular Piety 1550–1640* (Cambridge 1991).

4   See Hyder E. Rollins, 'The Black-Letter Broadside Ballad', *PMLA* 34 (1919) 258–339; Louis B. Wright, 'Ephemeral Reading', in *Middle-Class Culture in Elizabethan England* (London, 1936) 418–64; Bernard Capp, 'Popular Literature', in *Popular Culture In Seventeenth-Century England*, ed. Barry Reay (London, 1985) 198–243.

5   All 'default' references are to W.G. Day's five-volume facsimile edition of *The Pepys Ballads*, in the series 'Catalogue of the Pepys Library at Magdalene College, Cambridge' (Cambridge, 1987).

6   See, for instance, Edward Arber, *A Transcript of the Registers of the Company of Stationers of London, 1554–1640* (London, 1875–94) and Narcissus Luttrell, *A Brief Historical Relation of the State Affairs from September 1678 to April 1714* (Oxford, 1857).

7   This term was coined by Gérard Genette, in his *Figures III* (Paris, 1972).

8   Cf. Delany, pp. 78–9.

9   See for instance *The Pepys Ballads*, ed. Hyder E. Rollins, 8 vols. (Cambridge, 1929–32) III, 126, 138–9; VI, 86. Hereafter abbreviated *PB*.

10   *A Pepysian Garland; Black-Letter Broadside Ballads of the Years 1595–1639*, ed. Hyder E. Rollins (Cambridge, 1922) p. 288.

11   James Shirley, *The Court Secret*, ed. W. Gifford (1833) V, 1; v. 500, quoted by Natascha Würzbach in her *The Rise of the English Street Ballad, 1550–1650* (Cambridge, 1990) p. 278.

12   See *PB* VII, 83–4.

13   See *PB* VII, 181.

14   According to Carole Rose Livingston, in Chapter 4: 'Broadside Treasons and Ballad Treacheries', of her *British Broadside Ballads of the Sixteenth Century* (New York, 1991) p. 855, 'pleasing the crown was indeed the intention of virtually every ballad printed' in the century covered by her study. The Commonwealth interval with its perils to royalty was not likely to make Charles II a more liberal ruler in this respect.

15   *The Roxburghe Ballads*, 7 vols., ed. W. Chappell et al. (Ballad Society, 1872–1895; reprinted New York: AMS Press, 1966) IV, 121–9; see especially p. 127, footnote. Hereafter abbreviated *RB*.

16   Cited in *RB* V, 325, note.

17   Würzbach, pp. 163–87.

18   Claude M. Simpson, *The British Broadside Ballad and its Music* (New Brunswick, N.J., 1966) p. xi.

19 The texts of ballads about executions are sometimes referred to by scholars as 'Hanging Verses'; the term includes heterodiegetic accounts as well as 'autobiographical' ones, and may refer to executions of various kinds.

20 William Chappell, *The Ballad Literature and Popular Music of the Olden Time* (London, 1859; reprinted New York, 1965) p. 175.

21 Chappell, p. 176; Simpson, pp. 206–7.

22 I omit reference to ballads exclusively appearing in so-called 'Garlands', which were not sold as broadsides.

23 Francis James Child, *The English and Scottish Popular Ballads* (New York, 1884–98; reprinted New York, 1957).

24 The term is used by Shakespeare's Falstaff in *Henry IV Part Two* (III, ii) which appeared in Quarto in 1600; it refers to a kind of song or tune by then long out of fashion.

25 Cf.: 'In the popular ballads the Last Goodnight has the more poetic purpose of allowing a brave man to give a final display of courage. Such is the case in "Hobie Noble" (189) and in "Johnie Armstrong" (169B) where the Border hero bids a defiant goodbye to life...'; Alan Bold, *The Ballad* (London, 1979) p. 38.

26 *RB* gives the text of 'Good Night, and God be with you all; Or, The Neighbour's farewell to his friends', and dates it 'circâ 1750' (p. 312); this song is courteous and high-principled in tone; it has no associations with crime or punishment.

27 Examples include Dunbar's *Lament for the Makarys*, many of Chaucer's short 'Complaints', Lydgate's *Complaint of the Black Knight*; and Hoccleve's autobiographical *Complaint*; in the sixteenth century, the Complaint was the stock-in-trade of the love-poet, as Tottell's *Miscellany* amply illustrates.

28 See D.R. Woolf, 'The "Common Voice": History, Folklore, and Oral Tradition in Early Modern England', *Past & Present* 120 (1988) 36–7.

29 This interpretation is not contradicted but confirmed by 'King Charles, His Speech, and last Farewell to the World, made upon the Scaffold at Whitehall-gate, on Tuesday, January 30, 1648[–49]...' (*RB* VIII, xc***).

30 For examples, see the 'Appendix' in Würzbach, 253–84.

31 Rollins 1922, p. 425.

32 BM C.20.f.2, cited by Simpson, p. 748.

33 *RB* VIII, xc***, cited by Simpson, p. 748.

34 Rollins, p. 425.

35 See Simpson, p. 748.

36 Chappell, p. 163; also *RB* V, 558; *PB* III, 172.

37 Rollins, p. 278, footnote 40.

38 Henry Chettle, in *Kind-Harts Dreame. Conteining Fiue Apparitions with their Inuectives against Abuses Raigning* (1592), refers to ballad-writers as 'idiots that think themselues artists because they can English an obligation, or write a true straffe to the tune of fortune'.

39 In his communications to *Notes and Queries* on the Stationer's Registers, J.P. Collier claims the tune assigned to 'The Lamentacon of Christofer Tomlinson' (of 1592) is "Fortune"'; quoted by Francis Oscar Mann, *The Works of Thomas Deloney* (Oxford, 1912) p. 503. However, the first verses, which Collier quotes, do not fit that tune. The same objection applies to 'A New Scotch Ballad called Bothwell-Bridge' of 1679 (*RB* 4, 537–40).

40 'The original date was probably circâ 1590–1594, and it was entered for transfer on 14th December, 1624, as "George Sanders."', *RB* I, 75.

41 See also Rollins, p. 431.

42 *RB* VIII, 46–53; *RB* III, 144–6; *RB* III, 155–9 and *RB* III, 147–9; *RB* III, 150–4, respectively.

43 The only exception to the rule of sadness is 'A ioyful new ballad of the late victorye obtained by my Lord Mount Joy and our Maiesty's forces in Ireland', dated 1602 (*The Shirburn Ballads, 1585–1616*, ed. Andrew Clark (Oxford, 1907) XXXI, pp. 123–8).

44 *The Pack of Autolycus*, ed. Hyder Edward Rollins (Cambridge, Mass., 1927) 62–7.

45 Rollins, pp. 101–6.

46 Cf. Simpson, pp. 229–31.

47 II, 153; II, 169; II, 200; *RB* 4, 130–2.

48 See Simpson, p. 622.

49 They are, in Volume II: 161, 179, 183, 186, 188, 192, 194, 199, 237, 243, 245; and in Volume V: 5, 7, 8, 9, 11, 13, 15, 18, 20, 22, 23.

50 *Origins of the Individual Self: Autobiography and Self-Identity in England, 1591–1791* (Cambridge, 1997), Chapter 7, pp. 162–201.

51 Lacqueur, 'Crowds, Carnival and the State . . .' (see also note 2).

52 Cf. Rollins, p. 281.

53 John Taylor the Water Poet, *The Travels of Twelve-Pence*; in *All the Works of John Taylor the Water Poet* (1630; reprinted in facsimile London, 1973), I, 66.

12
Slightly Different Meanings: Insanity, Language and the Self in Early Modern Autobiographical Pamphlets

Allan Ingram

'Slightly different meanings': the phrase is John Perceval's, a man who spent several periods in asylums in the 1830s, and for whom language and its imperatives constituted a major factor in his insanity. Words, according to Perceval, acquire slightly different meanings in madness. This essay is concerned with the representation of individual identity in two 'mad' pamphlets, one from the 1680s, by Hannah Allen, and one, by Samuel Bruckshaw, from the 1770s. Identity for both these writers takes on slightly different meanings in the light of acknowledged or denied insanity, though I shall be arguing that in the case of Allen they are meanings of which she is almost in control, while Bruckshaw in his writing is as robustly unaware of the possibility of ambiguities, of other interpretations, as he was with regard to the train of events that brought him to confinement in the first place. Indeed, his pamphlet can be considered as largely an attempt to reconstruct those events in all the solidity of their apparent singleness of meaning. Each writer has a story to tell and a point to the telling of it, but that point, inevitably, is itself impinged upon by different factors contingent upon the telling, and upon each writer's stance of authorship.

Issues of gender are significant here, with Allen immersed in a family context, herself a daughter transferred from mother to aunt to brother and finally to second husband, while Bruckshaw, the self-reliant man of business, struggles on alone through an endless series of hostile trans-actions. So too are forms, particularly the forms of expression each writer acknowledges in order to validate the meanings they aspire to present.

Authority for Allen is in the Bible and its commentators. Biblical forms enshrine absolute truth, and their significance cannot be questioned. From this arises, in large measure, her particular sense of isolation. For Bruckshaw, on the other hand, if the power of his writing rests on self-assertion, on the insistence on an uncompromising self-definition, the authority towards which he leans is legal. Where Allen reaches for biblical quotation, Bruckshaw cites the forensic evidence. Where Allen requires total reabsorption into the certainty of a religious community, Bruckshaw looks for a favourable judgment that will endorse personality and his interpretation of events, preferably with exemplary damages.

Both these pamphlets are first-person narratives, and neither is mediated through any form of medical intervention. Indeed, medicine has very little part of either case, despite Bruckshaw's being confined as insane in a much more intrusively medical age than Allen's. The only sustained medical treatment Hannah Allen receives is from a nonconformist minister. However, the absence of a medical voice, perversely, liberates the autobiographical impulse to seek subtler ways of mediating the experiences related and of presenting identities, therefore, through the slightly different perspectives engaged by words under pressure from madness.

Hannah Allen's pamphlet *A Narrative of God's Gracious Dealings With that Choice Christian Mrs. Hannah Allen* (also known as *Satan's Methods and Malice Baffled*) was written in 1681 and published in 1683, but deals with events of the mid-1660s. Hannah Allen, as she tells us, was born Hannah Archer in about 1638 'of Religious Parents' (p. 5)[1] in Snelston, Derbyshire. Her father died while she was still young, and she was brought up first by her mother, apparently in comfortable circumstances, and from 1650 by relations in London. During this time she found that 'it pleased God to work in me earnest breathings after the ways of God', but also, more ominously, she became aware of the capacity of 'the enemy of my Soul' to 'cast in horrible blasphemous thoughts and injections into my mind, insomuch that I was seldom free day or night' (p. 5). At length, 'these Temptations grew to that height, that I was perswaded I had sinned the Unpardonable Sin' (p. 5). At this period it was the books of the early Puritan commentator Richard Bolton that enabled her to recover from 'my Despairing condition' (p. 6), so much so that in around 1655 she was 'disposed of in Marriage to Mr. *Hannibal Allen*', a merchant who travelled much overseas, to whom she bore one son (p. 6). It was Allen's death some eight years later 'beyond Sea' (p. 6) that brought about a recurrence of her 'deep Melancholy': 'no sooner', says Allen, 'did this black humour begin to darken my Soul, but the

Devil set on with his former Temptations' (p. 6). She gradually came to think herself abandoned by God, and therefore that she would be damned in perpetuity: 'Now little to be heard from me, but lamenting my woful state, in very sad and dreadful Expressions; As that I was undone for ever; that I was worse than *Cain* or *Judas*; that now the Devil had overcome me irrecoverably; that this was what he had been aiming at all along' (p. 9). She believed first that she would instantly die:

> One night as I was sitting by the fire, all of a sudden I said I should die presently; whereupon my Aunt was called; to whom I said, *Aunt, I am just dying, I cannot live an hour if there were no more in the world*; in this opinion I continued a great while, every morning saying, *I should dye before night*, and every night, *before morning*. (p. 9)

Then, as suddenly, she believed that she should kill herself:

> ...and now my opinion of Dying suddenly began to leave me, therefore I concluded that God would not suffer me to dye a natural death; but that I should commit some fearful abomination, and so be put to some horrible death:...to prevent which I studied several ways to make away my self. (p. 11)

Failing in her efforts to obtain opium, which, she felt, would enable her to die in her sleep, she took instead to smoking spiders in pipes of tobacco, though generally 'my heart would fail me' (p. 11) before the crucial moment. Finally, she decided that by eating she 'encreased the Fire within me, which would at last burn me up', and therefore she determined to starve herself in order, perversely, to stay alive, 'for I would now willingly live out of Hell as long as I could' (p. 18). She was never committed to a madhouse, and indeed her condition seems to have remained unknown outside her immediate family and neighbourhood. She was eventually restored to her senses, and to a belief in salvation, by the influence of her friend and relation by marriage, John Shorthose, a minister, and his wife:

> ...it pleased God by Mr. *Shorthose*'s means to do me much good both in Soul and Body; he had some skill in Physick himself, and also consulted with Physicians about me; he kept me to a course of Physick most part of the summer, except when the great heat of the Weather prevented, I began much to leave my dreadful expressions concerning my condition, and was present with them at duty;

and at last they prevailed with me to go with them to the publick Ordinance, and to walk with them to visit Friends, and was much alter'd for the better. (p. 19)

The improvement began in April 1666 and was completed by spring 1668, when she married 'one Mr. *Charles Hatt,* a Widdower living in *Warwickshire;* . . . my husband being one that truly fears God' (p. 19).

The pattern of Allen's pamphlet is conventional, and as a narrative it fits comfortably into the Puritan form of spiritual autobiography, that which A.W. Brink, in his edition of the nonconformist minister George Trosse's own relation of his life, refers to as 'moving accounts of guilt-ridden collapse and regeneration'.[2] These are texts that typically deal with temptation away from the love of God by the wiles of Satan, descent into mental and spiritual despair accompanied by intense physical suffering, often self-inflicted, and eventual restoration, from which perspective the tale is told for the benefit of fellow pilgrims. What makes Allen's writing distinctive is the individual form her madness takes, and her relation to the language in which it is represented. Such language is, of course, resolutely biblical, with quotation and allusion being habitually admitted as benchmarks for proper understanding. This serves both to standardise and to distance the experiences and feelings of twenty years earlier:

> One night, I said there was a great clap of Thunder like the shot of a Piece of Ordnance, came down directly over my Bed; and after that the same night, a while after, I heard like the voice of two Young Men singing in the Yard, over against my Chamber; which I said were Devils in the likeness of Men, singing for joy that they had overcome me; and in the morning as I was going to rise, that Scripture in the 10th. of Heb. and the last words of the 26th. Verse, was suggested to me from Heaven (as I thought) *There remains no more Sacrifice for sin;* And this delusion remained with me as an Oracle all along; that by this miracle of the Thunder, and the Voice and the Scripture, God revealed to me that I was damned. (p. 9)

If it is the glimpses of a normality observed as continuing just beyond the reach of her delusions – 'the voice of two Young Men singing in the Yard' – that are responsible for much of the refreshingly individual tone of Allen's writing, it is nevertheless the parenthetical '(as I thought)' that is particularly characteristic of the representational strategies of the pamphlet: she is quite consciously drawing a line between the self then and the self now, the significance of events as experienced and their significance from the security of a restored faith in salvation – 'the Triumphant Victories, Rich and Sovereign Graces' (p. 1), as she announces on the title-page. That same line is also drawn with regard to names. The Hannah Allen who writes is in fact Hannah Hatt – '(Afterwards Married to Mr. *Hatt*)' (p. 1), again, parenthetically, on the title-page. Prior to that she was Hannah Archer. Each name conveys different meanings: the confirmed Christian whose husband 'truly fears God' (p. 19); the daughter who, while pre-empting some of the anxieties of her young adult self, was raised 'in the fear of God from my Childhood' (p. 5); and Hannah Allen whose experience of melancholy, despair, self-loathing and attempted suicide is to be held up as an example to all good Christians. Only Hannah Allen was fallible enough to suffer prolonged temptation by the devil, to fall prey to doubt over herself and over the Almighty.

This tripartite identity is glossed over by the title's triumphant asser-tion, *A Narrative of God's Gracious Dealings With that Choice Christian Mrs. Hannah Allen*, yet the grace of God has in fact been enjoyed by Hannah Hatt, while being 'chosen' for Hannah Allen meant individual experience of the most negative and self-destructive kind:

> Many places of Scripture I would repeat with much terrour, applying them to my self; as Jer. vi. 29, 30. *The bellows are burnt, the lead is consumed of the fire; the Founder melteth in vain; Reprobate silver shall men call them, because the Lord hath rejected them*; Ezek. xxiv. 13. *In thy filthiness is lewdness, because I have purged thee and thou wast not purged; thou shalt not be purged from thy filthiness any more, till I have caused my fury to rest upon thee*: Luke xiii. 24. *Strive to enter in at the strait gate, for many I say unto you, will seek to enter in and shall not be able.* This last Scripture I would express with much passionate weeping, saying, *This is a dreadful Scripture, I sought, but not in a right way; for the Devil blinded mine eyes, I sought to enter but was not able.* (p. 10)

The relation between the biblical and the individual is sharply scored here, with repetition of a sequence of already familiar phrases pro-ducing the unfamiliar 'terrour' of self-realisation, and for Allen the lin-guistic aspect of that relation is particularly problematic, not least in her inability to discover a sufficiently terrible biblical exemplar for her to make final sense of her own condition, obliging her instead to mod-ify and extend what has been scripturally handed down:

I would often say, I was a thousand times worse than the Devil, for the Devil had never committed such Sins as I had; for I had committed worse Sins than the Sin against the Holy-Ghost: some would answer, *The Scripture speaks not of worse sins, and can you be guilty of greater Sins than the Scripture mentions?* Yes, said *I, My Sins are so great, that if all the Sins of all the Devils and Damned in Hell, and all the Reprobates on Earth were comprehended in one man; mine are greater; There is no word comes so near the comprehension of the dreadfulness of my Condition; as that, I am the Monster of the Creation*; in this word *I* much delighted. (p. 13)

Identity for Hannah Allen can be comprehended only in terms of biblically validated labels, or rather of biblically unvalidated silences (of 'greater Sins than the Scripture mentions'), realised for the first time through the horror of her own experience. Identity for Hannah Hatt, however, is to be able to see that experience from the other side of a line.

At the same time, Allen's very 'delight' in the phrase she has found to describe herself, '*Monster of the Creation*', represents a slightly different emphasis from her biblical models, a departure in the direction of individuality. Her confessed delight affords a degree of insight into the feelings of a woman who was finding that the resources of language, and especially of language sanctified by scripture, were closed to her as an expressive medium. The language in which Allen had once put faith had nothing to say to her, apart from a blanket rejection. It could only condemn. It made no acknowledgement of her condition, left her to find identity in its silences, in what it did not express.

Banished to the fringes of language, she manifests, too, an urge for physical as well as linguistic seclusion. She shuns the visits of friends. She resents being prayed for, saying '*they did not pray for me, for I was not to be prayed for*; for the Scripture said, *That they who had sin'd the sin unto death, were not to be Prayed for*' (p. 13). She fears travelling from her home in Derbyshire to London in company: '*I would earnestly argue against it, and say, I shall surely dye by the way, . . . Mother, do you think people will like to have a dead Corps in the Coach with them?*' (p. 10). This urge is particularly powerful in relation to her drive towards self-destruction – her desire '*to get into a Wood and dye there*' (p. 14), and her attempt while at her cousin Walker's to hide herself away and starve to death:

. . . the Devil found me out a place on the top of the House, a hole where some boardswere laid, and there I crowded in my self, and laid a long black Scarf upon me, and put the boards as well as I could to hide me from being found, and there intended to lye till I should starve to death; ... but when I had lain there almost three days, I was so hungry and cold, it being a very sharp Season, that I was forced to call as loud as I could, and so was heard and released from that place. (p. 12)

The hidden self of Hannah Allen, 'crowded' under the floor boards or finding only the occasional word that would comprehend her sense of her own monstrosity, is slightly different from the 'Choice Christian' offered to the believing world by Hannah Hatt. This difference becomes especially focused at a crucial moment of narrative silence, a point where the individual and the biblical in fact would seem likely revealingly to mesh. In the event, which is the single unquoted biblical reference of the whole pamphlet, it is the silence that reveals. In a text in which biblical quotation and biblical allusion are rampant, one verse only remains hidden under a simple reference: '*Col.* 3.5.' (p. 13) What could have been fully quoted is St Paul's uncompromising declaration: 'Mortify therefore your members which are upon the earth; fornication, uncleanness, inordinate affection, evil concupiscence, and covetousness, which is idolatry.' Here at last, in this act of self-censorship, of conscious half-concealment, is an indication of what is driving and at the same time being suppressed by the images of self-loathing, a hint of the emotions that drove the young Hannah Allen into despair, made her see herself as a guilty, worthless creature, as 'the Monster of the Creation'. Beneath the representations, first as monster, and then laid over the monster the 'Choice Christain', there persists an undeniable imperative. It is sexual and it is human, but it is also 'unclean' and 'inordinate', and associated perhaps with the absent and rarely mentioned Hannibal Allen, dead 'beyond Sea'. Unspoken, apparently unspeakable, hidden at the very edge of self-representation, as deeply as language can hide it, the impulse remains, crying out to be found under the boards of biblical correctness, a key to the Hannah Allen that Hannah Hatt could not forget, but could not wholly acknowledge either.

Acknowledgement was no problem for Samuel Bruckshaw, largely because he could see nothing that he had done wrong, certainly nothing to deserve confinement in Wilson's private madhouse in Ashton-under-Lyne for nine months between June 1770 and March 1771. Bruckshaw was a wool-stapler, buying in, grading and selling on wool, at Bourne in Lincolnshire, having moved there from Cheshire, from where his family originated. In 1768 he purchased a new property for his business in nearby Stamford at a more profitable location 'upon the North road' (p. 80),[3] which both overstretched his resources and aroused local resentment against him. A calling in by his creditors of outstanding debts meant that he straight away lost possession of the new premises to Mr John Bowis, attorney-at-law, who immediately disposed of them to a Mr Langton. An angry visit by Bruckshaw to Langton at the house in question set in motion a train of events that included his arrest for threatening behaviour, his being beaten and robbed by his gaolers, a harrowing journey over three days by post-chaise to Aston-under-Lyne with the sinister, half-comic Wilsons, and his detention in their madhouse:

> When Wilson shewed me to bed, he carried me up into a dark and dirty garret, *there stripped me*, and carried my cloaths out of the room, which I saw no more, *for upwards of a month*, but lay chained to this bad bed, *all that time*; this appears to be their breaking in garret; under the ridgetree is a box for the harbour of pigeons, which they disturb in the night time, to affright their prisoner when he should rest. For this purpose some of Wilson's family are up all night long, sometimes they throw pails of water down under the window, now and then brushing across, with a few small rods, or rubbing with a stone or brick upon the wall, sometimes put a light up to the window, and every now and then make a disagreeable noise, to awake you in a fright. In the day-time the window is darkened, and common necessaries denied; they gave me bad victuals, short allowance, with sour beer, oftener water, and sometimes not that; no attendance, but what was as contradictory and provoking as they could possibly invent, and frequently the most barbarous stripes.... Wilson's wife threatened me thus: *'I'll lay yoo o' the head with the poker, if yoo do make complaint to any body that comes into th' house.'* (pp. 95–6)

While there, Bruckshaw was periodically kept in irons, threatened, confined in a smoky garret, deprived at times of sleep, warmth and food, and refused visitors until released by the agency of his brother Joshua. He thereupon transferred to London and set about attempting to regain his business credibility, without success. He also began, through long and costly legal proceedings, to pursue financial redress from John Hopkins, Mayor of Stamford, and John Exton, alderman, whom, as magistrates, he held responsible for the original order of confinement. His failure, with attorney after attorney first accepting and then withdrawing from his case, with two trials bungled by his representatives and dismissed as 'nonsuits', and with the two defendants successfully seeking immunity by virtue of their office, confirmed him in the view that there was a legal conspiracy against him. A last desperate resource was a petition to the monarch at Christmas 1773 and, after that inevitable rejection, the writing and publication in 1774 of *One More Proof of The Iniquitous Abuse of Private Madhouses*. The fact that this was 'Printed for the AUTHOR, and to be had of him, at No. 28, Poultry' (p. 75) is an indication of Bruckshaw's utter isolation, even if it is also a measure of his undeniable resilience. A second pamphlet, *The Case, Petition, and Address of Samuel Bruckshaw*, appeared in the same year.

Bruckshaw's case is almost the exact reverse of Hannah Allen's. Where she acknowledged the madness of her former behaviour, and represented it almost entirely through the language and modes of an acceptable religious interpretation, he finds himself, in a context in which religion makes few if any calls, in a position of having to prove his own sanity, a position in which all the signs a writer can make are the wrong ones. Where Allen could draw a line between narrator and victim, Bruckshaw the protester and Bruckshaw the imputed lunatic are one and the same. The generation, therefore, of meanings that are slightly different from those intended is legion.

No reader of this pamphlet can doubt that Bruckshaw was very badly treated, especially at the Wilson's disgraceful madhouse, but equally a minor adjustment in perspective gives a view of events in which he has himself behaved in ways likely, in a provincial town in the late eighteenth century, to pass fairly convincingly for mad. After his first visit to Langton, for example, he receives a copy of a message to the mayor alleging a second visit that had so upset Mrs Langton, who was with child, that she 'is like to miscarry' (p. 87). Bruckshaw immediately returns to 'ease her of her fears', and finds 'two constables placed with their staffs within the yard gates'. Upon Langton's command, 'Ely Buswell, seized me by the collar: in return, I seized Buswell by his, demanding his authority, who up with his staff to knock me down, which I catched in my other hand; then Needham, the other constable, collared me, and they dragged me through the public streets... to the Mayor's'. Mrs Langton and the dangers of her miscarrying are forgotten while Bruckshaw, fitting another piece into his conspiracy theory, concludes that the message 'was written with a design to send me irritated against Langton ... as the constables were placed there ready for seizing me' (p. 87). He suspects, too, that the medicine prescribed for him while in gaol by Dr Jackson, who has attended at Bruckshaw's own request, is poisoned, and intends to 'have it examined by the college of physicians' (p. 89) until it is stolen from him by his gaolers. Even, in his increasing isolation, the few signs of local support that he presents are ones that tend to confirm the townspeople's view of him, particularly the mysterious voices: 'In the dusk of the evening a male voice, seemingly desirous to be unknown, in passing under the town-hall, called out, "Continue to behave like a man, you will have some desperate work presently"' (p. 90); and, from the garret at the gaoler's house, under attack from gaoler Clarke and his servants Walker, Whittle and White,

> I then threw open the window, and called out, as loud as I was able, 'Murder! murder! murder!' in hopes that some humane person or Peace Officer would come to me and enquire into the reason of my cries, but I was answered from the bridge, 'You may call out, but they will let nobody come to your relief.' (p. 91)

Bruckshaw is disastrously within the skin of his own narrative, seeing now as he saw then and locked therefore within the signs and causes of his present anger and despair. 'Upon my enlargement,' he says,

> I went amongst my acquaintance, and found many of them irremovably prejudiced, with a belief that insanity was the cause of my imprisonment at Stamford; which filled my mind with much anxiety and displeasure against my unjust oppressors, for having so infamously and irreparably injured me. (p. 98)

Yet the damage to his business connections is equalled by the harm done to his narrative interests. Reading and understanding of Bruckshaw's self-published, self-distributed pamphlet are conducted through the perspective of the madhouse, even though the madhouse was as disreputable and inadequate as Wilson's. (It was officially licensed in 1780.[4]) Nothing in the narrative actually proves that Bruckshaw was mad, but his whole enterprise exemplifies the linguistic difficulties involved in demonstrating oneself sane.

Psychological and linguistic isolation become more and more apparent as the pamphlet moves into its desperate second half, with Bruckshaw tramping through London's legal networks, putting himself into the hands of its agents.

> Then Mr. Crispin took this opportunity of declining prosecuting my action any further; after thus trifling with me till it was out of

my power to get properly fixed with another ... therefore I took my affairs out of his hands .... At length a friend of mine informed me, that he knew a Mr. Jenkenson, Attorney at Law, in Hoxton Square, who had in several instances gone through business in his profession with great integrity, after others of reputed abilities, &c. had thrown cold water upon the matter; whereupon I laid my business and papers before him, who was dissatisfied with Mr. Wallace's aforesaid opinion, and laid it before Serjeant Walker; and upon the 30th of October, we went together to take his opinion, upon Mr. Wallace's opinion and case; when he informed us thus: 'I have looked over this case very carefully, but will read it over again.' then said, 'I am clearly of opinion, this action will lie.' Mr. Jenkenson then asked him, if he would advise him to go on with it; to which the Serjeant answered, that is no part of my business, that belongs to Mr. Bruckshaw. I have only to say, as I said before; 'that the action will lie.' We withdrew and parted in Fleet Street, when Mr. Jenkenson said, "I will give the Defendants notice of trail, as soon as you please." A few days after, I saw him, and the aforesaid Mr. William Bolton, walking arm in arm, along St. Paul's Church Yard, and the next time I called upon him in his office, he began to raise objections, which I interpreted into some desire he had to decline; therefore I told him, if he chose to give me my papers, we would part, which he immediately did. (p. 104)

Bruckshaw becomes increasingly absorbed by the procedures of legality and litigation, adopting its terms and technicalities to the further detriment of his narrative coherence. The man who could seize Ely Buswell's staff of office and demand to see his authority is floored by legal forms and intricacies. Robust narrative self-assertion gives way to the shorthand account of court proceedings at which Bruckshaw himself has been only a passive observer:

Halford Allan, sworn. Says 'he was servant to Langton;' and my counsel have taken his evidence down in these words: 'Plaintiff forced his way into Langton's house, and was near knocking the witness down. Langton sent him to Mayor Hopkins, to lay a complaint that Plaintiff was a madman, had broken into his premises, desiring him to send his peace-officers – peace-officers came immediately.'
His Cross-Examination taken by Mr. Balguy – *thus.*
Don't say Plaintiff touched him – no writing at the time of his applying to the Justice Defendant.
Defendant's second Witness taken by my Counsel in these Words.

> *Elias Buswell, sworn.* 'Was a constable at Stamford; June 1770, he
> had orders from the Town-serjeant by Thickbroom's orders – Plaintiff
> came in and behaved ill, struck him, and threw him against the
> wall.' . . .
> *Cross-Examination taken down by* Mr. Wheler. (p. 112)

Equally, Bruckshaw's attempts to consolidate reader support are based
on an increasingly fragile assumption of common forms of language.
'Fabrigas against Mostyn', he has his reader helpfully interject, referring
to a case of 14 Geo. III in which Fabrigas, a native of Minorca, successfully sued the British governor, Mostyn, in an English court for wrong-
ful confinement. Fabrigas was awarded damages of £3,000. 'True, my
good Friend,' returns Bruckshaw, looking in generous pity at the innocence
of the indisputably sane world, 'but Fabrigas was supported by a Noble
Duke, of princely spirit, and of princely fortune; and happy it was for
him that he gained that protection' (p. 98). (The reference is to Charles
Lennox, 3rd Duke of Richmond, who later supported the American col-
onists in parliament.) Similarly '24th Geo. II. cap. 44' becomes almost a
choric refrain as Bruckshaw falls increasingly under the spell of the Act
that, maddeningly, is affording protection to his persecutors – 'An Act
for the rendering Justices of the Peace more safe in the Execution of
their Office; and for indemnifying Constables and others acting in
Obedience to their Warrants', which became law in 1751. The pamphlet
ends with a last desperate cry of an isolated individual, for which Bruckshaw chooses to move, at least partly, into the third person, as if himself
addressing the court of universal justice which, he hopes, will rectify
the wrongs done him in the narrowly self-interested English courts:

> Upwards of Five hundred and twenty pounds has it cost this man in
> paying his own law expences and the costs of those who have thus
> destroyed him.
> 'To hope every thing even against hope.' –
> There is I think somewhere such a sentence, and the case is not
> imaginary; for this much injured man is yet willing to believe that by
> the aid of such of his fellow subjects as may think his case deserving
> their assistance, he may yet drag to the altars of the offended law,
> victims the most worthy ever sacrificed on them. (p. 115)

An appendix contains fourteen documents, including extracts from
letters, and sworn statements testifying to Bruckshaw's sanity:

WE whose names are hereunto set and subscribed (according to the dates affixedtherewith) did see and converse with Mr. Samuel Bruckshaw, and from the result of such conversation never had the least reason to believe him disordered in his mind. – Witness our several and respective names, this 30th day of April, 1771. (p. 125)

In the slippery enterprise of self-representation, one is only as sane as the words on the page. To 'see and converse' is to confirm one kind of truth, and belief can be effectively sworn in legal form. Writing beyond the law, on the other hand, leaves every slip of the pen open to interrogation, and for Bruckshaw writing was like an ice rink. Hannah Hatt wrote in confirmation of biblical authority, to celebrate the genuinely gracious nature of God's dealings with her former self, yet in doing so encountered a necessary silence that remains a revealing textual unspoken, a moment of truth beyond the Christian witness to which her pamphlet swears testimony. Bruckshaw wrote in confirmation of self, attempting to make coherent the experience of being treated as a lunatic, and replacing society's consequent reticence with protestation, evidence and assertion, forms of expression that confirm forcefully the wish to be believed but leave largely unspoken the validity of the identity that is to be confirmed. There is no record of Bruckshaw ever succeeding in his legal quest. Unlike Hannah Allen, he found it impossible to come in from the cold of imputed madness. Identity, for Bruckshaw, remained locked in the terms and conditions of the insanity he so vigorously resented, terms that readily encouraged different readings from Bruckshaw's own. Hannah Hatt was able to leave madness behind, calling it Hannah Allen, as she addressed her own community of Christian readers as a respectable married woman. Yet even in her case, as we have seen, the effort to express past identity under stress brought into play slightly different meanings, slightly different perspectives. In the context of both these narratives, their controlling interpretations and their underlying vulnerabilities, that point where slightly different meaning is engaged is also the point where self-representation as a self-conscious enterprise begins to break down. At that point, individual identity ceases to be a matter of signification and starts to become a matter of being.

Notes

1 All quotations are from *Voices of Madness: Four Pamphlets, 1683–1796*, ed. Allan Ingram (Stroud, 1997).

2   *The Life of the Reverend Mr. George Trosse, Late Minister of the Gospel in the City of Exon, Who Died January 11th, 1712/13*, ed. A.W. Brink (Montreal and London, 1974) p. 38.
3   All quotations are from *Voices of Madness: Four Pamphlets, 1683–1796*.
4   See Millicent Regan, *A Caring Society: A Study of Lunacy in Liverpool and South West Lancashire from 1650 to 1948* (Merseyside, 1986) p. 22, citing Manchester Quarter Sessions Petitions Order Book, 12 October 1780.

13
Epilogue: 'Oppression Makes a Wise Man Mad': the Suffering of the Self in Autobiographical Tradition

Elspeth Graham

My starting place is with two commonplace assumptions, one formal, the other historical. First, then, formally in autobiography we expect to find an intensity of focus on selfhood as a source of meaning, and conventionally see the defining features of the genre as lying in the coincidence in it of subject and object, and of enunciator, enunciation and enunciated (or narrator, narrative process and story). Then, historically, the emergence of autobiography as a significant genre is commonly seen in the early modern period as an aspect of the rise of individualism in western societies, although it is not named, and therefore not fully designated as a genre, until the beginning of the nineteenth century. So the autobiographical impulse, with all that it implies, and modernity are thus commonly seen to be inextricably linked.

Those central features of individualist thought – recognition of the individual as a source of meaning; the notion that the value of the individual lies in his or her own distinctiveness, particularity or uniqueness; the idea that the individual bears responsibility, if not for what happens to him or her in a life, then for the making sense of it, for the gathering of experience to the self in order for meaning to be made, in order to be one's own person – are not only central to western democratic systems, but are intrinsic to autobiography as a genre. Autobiography and individualism both imply some sort of dialectic between the agency of the individual and awareness of the self, or self-consciousness.

But the generic grouping of texts dependent on representation of the self as the prime source of meaning is, as the introduction to this book suggests, also complex and troublesome. There are, after all, many sorts of autobiographies: those of groups as well as those of individuals; those of the famous and those of people who are known simply because of their stories of themselves; those which record events of a life quite matter-of-factly – even cheerfully – and those which structure their representations of selves on moments of crisis (and perhaps recovery). This last sort of autobiography is the one which has been most influential in the development and production of autobiography from the early modern period to the present and which has received by far the greatest amount of attention in literary history and criticism. Our current preoccupation with autobiography, the prime genre of selfhood, very often seems to be, in fact, a preoccupation with threats to selfhood, with issues of collapse and absence of self, with the very fragility or impossibility of a concept and sense of the self. A quick scan through recently published autobiographies and the titles of academic papers analysing autobiography suggests clearly that issues of crisis and suffering are central to what we expect of, and are interested by in, autobiography. We find concentration on issues of: absence from the world, from history, in sickness and death; insanity, ecstasy and the sublime; the self presented through its deletion or erasure or disappearance or evasion, through the political oppressions of slavery, apartheid, gender, sexuality and class. Selfhood, it might seem from this, is not a given: it is produced through suffering and is knowable or representable only *in extremis*.

My concern in this chapter is with these issues of suffering and crisis as central to a dominant – if not the only – convention of autobiographical writing. The very simple perception that I want to work from concerns the gaps that open up between consciousness of self and agency, or that emerge from the representation of selfhood as simultaneously an affirmation of individuality and a statement of the fragility of self. If autobiography is concerned with the representation of self, it is equally, very often, bound up with lack of selfhood. I want, in particular, to investigate structuring patterns of suffering and absence in three ways: to consider suffering as it is psychically and formally linked to the very possibility of articulation of the self in autobiography; to trace briefly how social and political contradiction, conflict and contestation are important to the autobiographical enterprise; and to contemplate ways in which autobiography constitutes an affective dialogue between reader and autobiographical subject and how suffering may be seen as a cultural bonding force in individualist societies. And I will do this by looking briefly at two historically distant pieces of autobiographical writing – not in order to erase historical difference, but in order to show how the particularities of historical experience are mediated by the force of autobiographical form. Both *An Evil Cradling*, Brian Keenan's remarkable account of his kidnapping by fundamentalist Shi'ite militiamen in Beirut in 1985, and Evans and Cheevers' *A Short Relation of Cruel Sufferings*, the account of two Quaker women imprisoned by the Italian Inquisition in Malta in early 1660s, are stories of imprisonment, of threats to selfhood, and resistance to the forces which threaten the self. Both are narratives that depict physical and psychic suffering; both depict moments of ecstasy.

An Evil Cradling

What immediately strikes me as interesting about Brian Keenan's account is that it *is* presented as autobiography. The beginning of the first chapter of *An Evil Cradling* speaks of the need to find meaning for the experience of Keenan's kidnapping and imprisonment. The text that is produced out of this seeks to find that meaning by allocating responsibility for the experience to Keenan himself, by containing it within the structures of the self, by seeing its origins in the motivations and impulses of the self. The experience has to be possessed. What might otherwise be an act of violence – purely from without – demands to be integrated. The need for a source point of meaning, a beginning, is attached to selfhood. It is equally attached to words, his use of the biblical 'In the beginning was the word, and the Word was with God'[1] – a curiously unexplained reference – the need for words, and the need to see meaning, is aligned with the need to find a beginning. The need to understand then blurs into the need to share with the reader – to make 'you' understand. Indeed, the questions of the self, the demand for cause and origin, become the words of others: 'I have been asked so many times "Why did you go?".'[2] This seems an extraordinary question to me. It is as if the only way that we can comprehend an experience so threatening to selfhood and identity is to search for an act of personal agency on Keenan's part that explains it, that sets things in motion. Keenan and those others who have questioned him, and implicitly his readers, enter into an act of complicity right from the start of the text.

Keenan goes on to situate his account of imprisonment in a narrative that begins with his own restlessness, his own sense of entrapment in Belfast before he set out to the Lebanon. He locates his desire to travel in a series of confrontations with types of deaths: the actual death of his landlord as a student; his political reaction to Bloody Sunday in Belfast and his sense of entrapment by the political situation in Northern Ireland; and his more generalised fear of attachment (of marriage, of settling down) and the immobility he fears in that. His actual kidnapping is thereby located as an enactment of psychic and political experiences and configurations which are already a part of him. The imprisonment in Beirut is almost established as an ironic metaphor for an existing sense of selfhood.

One aspect of his extraordinary account of imprisonment which particularly interests me in this context is his description of several periods when he experienced ecstasy. Keenan describes times when he experienced hallucinatory visions, delirium and delusion and sometimes, transcending these, ecstasy. At one moment he describes how he hears imaginary music and how, in a seeming trance, he begins to dance, then dances on and on, naked, sweating and ecstatic:

> it seemed that I would be dancing forever.
> ...I had looked upon myself enraptured in this primal dance. I had seen myself go with this moment of ecstatic madness and had come back from it unmarked. (pp. 79–80)

Often, in his account, hunger, food, delirium and ecstasy are connected, as when he is enraptured by a bowl of oranges:

> I cannot, I will not eat this fruit. I sit in quiet joy, so complete, beyond the meaning of joy. My soul finds its own completeness in that bowl of colour...Everything meeting in a moment of colour and form, my rapture is no longer an abstract euphoria. (pp. 68–9)

At another time he fasts, becomes weak and light-headed:

> They were probably worried that I was already becoming ill. They explained that they had no doctor. I answered that I did not want a doctor. If they brought me one I would refuse to see or speak with him.

The refusal of food gives him a rhapsodic sense of power:

> I was confident, I was strong-willed and almost ecstatic as I pushed each meal from me ... I was beyond desire ... Here was a game I was winning; I was in control and control could not be taken from me.[3]

Elsewhere, he ruminates on his gaolers' 'moronic and ecstatic chanting' of the suras of the Koran, implicitly distinguishing this 'holy insanity' from his own range of psychic responses to imprisonment – his emotional shifts, intoxications and hallucinations, phases of apparent 'religious fervour', outbursts of abuse and profanity.[4]

How do we understand this variety of experiences which, along with numerous other states, Keenan labels as 'ecstatic' or 'euphoric'? Keenan himself provides several speculative interpretations of such states. He applies psychological explanations, suggesting that 'denial' of 'traumatic transitions . . . stimulates a euphoric state' (p. 31). And he is culturally and politically self-conscious in his exploration of the meanings of his refusal to eat, relating this to the history of Irish hunger-striking, which

> overcomes fear in its deepest sense. It removes and makes negligible the threat of punishment. It powerfully commits back into the hunger-striker's own hands the full sanction of his own life and of his own will. (p. 55)

I will return to ways in which we might read these experiences later. The point I want to consider here, however, concerns our reaction as readers to what is an horrific account. *An Evil Cradling*, along with the related accounts of Keenan's co-hostages, are best-sellers in Britain. The compulsion to read such narratives derives, clearly, from complex motivations. Although there will be different understandings of the politics of the texts according to our own alignments and identifications and the differences between the texts themselves, these narratives all, in some way, confirm the resilience of familiar values: humanity, love, the endurance of the self under extreme threat.[5] But, paradoxically, amongst the appeals of such narratives, there are also a cluster of compulsions which arise out of a sense of the unimaginability of the experiences which Keenan so powerfully describes or which emerge from the highlighting of collective cultural anxieties. Keenan's situation itself touches, I suspect, on common terrors. At one level, the arbitrariness of the act of kidnapping foregrounds the dynamics of terrorism itself. It brings to the surface the tension that exists in a Western culture that is highly individualist (valuing the coherence, uniqueness and agency of the individual) but which is equally a mass culture, dominated by the global mass media which makes the act of terrorism possible, and where the individual is 'anonymous and perfectly undifferentiated'.[6] Our vast uncertainties over the nature and status of individual subjectivity are highlighted. Fear and fascination greet the gap between versions of the self which represent, on one hand, the individual as all, as the source of

all meanings, and on the other, the individual as faceless, entirely inter-
changeable with nameless others. *An Evil Cradling* embodies a battle
between such versions of self.

At another level, we identify more simply with the individual as victim
and recognise that to be held in such a form of captivity represents an
ultimate test of being – of identity, selfhood and self-possession. Our
notions of what it is to be a person are examined in extremity.[7] The
moments of near-disintegration of selfhood which are experienced,
speak to our dread, but also fascinate us. In a late twentieth-century
society where deep anxiety prevails about the possibility of asserting
shared values and where the notion of identity itself is under threat, a
narrative that describes manifestations of subjectivity under duress
through the form of an inherited autobiographical structure (in which
self is traditionally put under threat, in which values are violently challenged but where personal integrity prevails, allowing for an affirmation of joy and human relationship) may be compelling. By inserting
unimaginable experience within a framework of recognisable autobiographical development, cause, suffering and threat to the integrity of
selfhood, Keenan may not only be seeking for a way to integrate his
experience, make it his own, or in the words of John Bunyan, the seventeenth-century nonconformist, bind it to him, but he also allows for a
recognisable structure of suffering which we as readers can share.[8]

Keenan's autobiographical account ultimately derives its structure
from its seventeenth-century antecedents: the often harrowing accounts
of spiritual and physical suffering, sometimes involving imprisonment or
fantasies of imprisonment, of Puritan and nonconformist confessors.[9]
Both Keenan and seventeenth-century spiritual autobiographers describe
subjectivities in crisis. These crises, as experienced subjectively but as
offered to readers to share and empathise with, also point to how the
writings of seventeenth- and twentieth-century autobiographers alike
may also be seen to instance different crisis moments in the history of
individualist belief. For all the crucial differences between Keenan's
experiences and those of seventeenth-century confessors, there are also
formal links and similarities in content between their self-narratives.

Evans and Cheevers

This cluster of issues I have extracted from *An Evil Cradling* – imprisonment, fasting and ecstasy – provides my link to my second text: Evans'
and Cheevers' *A Short Relation of Cruel Sufferings*. Quakers in the seventeenth century, in contradistinction to Calvinists, belonged to what was

'essentially an ecstatic movement'.[10] Unlike Calvinist nonconformists in the seventeenth century, Quakers believed in the availability of salvation to all who recognised Christ within them. Christ, the Inner Spirit or Light, was the prime guide to Quakers, more important than the Scriptures. God worked through the living Word. This indicates Quaker distrust of reified doctrinal interpretations of the Word of God, but can be further linked with Quakers' tendency to identify themselves as the incarnation of the Word, as living texts. The notion of Christ within, along with the concept of the living Word, led easily into the millennial notion that Christ had actually come in the form of Quakers, and even led to their direct identification with Christ. The Inner Light established unmediated rapport between individuals and between individuals and God, or even, it seems, led to a merging of identities: this merging of self in Other crucially marks the ecstatic nature of their belief and experience. The immanence of Christ in the lived being of Quakers might suggest, too, that a key biblical origin lies, for them, not only in the recognition that 'In the beginning was the Word', but in the follow-up, 'The Word was made Flesh'.[11] Words and bodies are conflated for Quakers.

Before the 1670s Quaker doctrine was not codified, and in the aggressive, strongly evangelical phase of Quakerism in the 1650s and 1660s Quaker belief was often defined 'defensively, by negatives', that is, in defence against attribution of beliefs by their opponents.[12] We may also see in this, however, a characteristic common to oppositional groups (and Quakers were, so to speak, doubly oppositional): the need to define self and belief through statement of what it is not – whether the beliefs of an opposed orthodoxy, or of opponents' definitions of the group. This oppositional element of self-definition may also be seen in the extent to which Quakers attracted hostility. Their aggressive and confrontational stance encouraged strong response. Much has been written about ways in which Quakers were perceived as a political threat, and how the law was used against them, but what has been less commented on is the extent to which Quakers seem to have courted hostility and why this was necessary to their mission.[13] This is not to suggest that they were complicit, in any pejorative sense, in their oppression. Rather, it is to remark on the necessity of undergoing oppression as formative of and intrinsic to their spiritual position.

To demonstrate the dynamics of this I would like to focus on that part of *A Short Relation of Cruel Sufferings* which describes Evans' fast and the surrounding debates concerning her expected death.[14] This provides, of course, a point of direct comparison between Evans and Cheevers on one hand, and Keenan on the other: fasting and attendant moments of ecstasy are a shared response to imprisonment. But unlike Keenan, Evans and Cheevers took on a directly antagonistic role: they set out to convert, and their strategies in the Inquisition were not merely those of resistance, but were equally motivated by their continuing desire to assert the truth of their beliefs, to convince and convert the Friars as much as to reject the Friars' counter-attempts to convert them to Catholicism. Their stated need to establish a means of survival is more ambiguous than Keenan's; they tell us later: 'We sought death but could not find it. We desired to die, but death fled from us' (p. 124). And it is their negotiation of fasting, ecstasy and the desire to control the meanings of death that interest me in this part of the text.

At this point in the narrative, Katharine Evans is ten days into her fast and apparently near to death. Fasting was a common Quaker activity: James Naylor, George Fox, Richard Hubberthorne and James Parnell (who died as a consequence of fasting in prison) were among the many who fasted. Such fasts involve identification with Christ (established in the passage from Evans and Cheevers by the reference to Christ's entombment after his crucifixion) and are in turn linked to notions of perfectibility and the elevation of the perfected spirit above the mortal and bodily.[15] In Evans' case, however, fasting is also linked specifically with the circumstances of her and Cheevers' imprisonment. Diane Purkiss suggests that this fast was intended 'not to influence their captors, but to influence God'.[16] I do not entirely agree. At one level we can easily see Evans' self-starvation as a response to the predicament caused by its being Lent.[17] To submit to the dietary restrictions demanded of her and Cheevers in their imposed Catholic environment would be to yield, however unwillingly, to the demands of oppositional doctrinal practices. When the Friars subsequently offer Evans, in her weakened state, a dispensation to eat meat, this is equally unacceptable: it would again imply her complicity in Catholic doctrine. To eat at all in the context of the doctrinal legislation of consumption is to collude in those doctrines. But beyond this we may see further motivations to her refusal.

Imprisonment itself may be seen as an epitomisation and condensation of the conflict between different Christianities, or more generally in the case of political or terrorist imprisonings, between different cultures and political positions. In imprisonment the realm of conflict is reduced to the confines of a cell, where the religious or political or cultural agon is played out with telescoped intensity, and where a battle of cultures becomes a battle waged at the level of the reconstitution, preservation or destruction of the subjectivity – bodily or psychic – of the prisoner.[18] (It has been argued, moreover, that the body, or in modern individualist societies, the unified self, is psychically equated to a prison or cell: both mark out the boundaries of the subject as distinct from a surrounding world. The shift of action from the actual place of the prison to the site of the prisoner's subjectivity is made all the more inevitable by such an equation.[19] We might also consider the prison a liminal region, prefiguring the tomb, Marvell's 'fine and private place', the locus of both the ultimate mutability of the body and of final self-definition. And, as Foucault points out, in speaking of confining institutions generally, 'the dormitory is the image of the sepulchre'.[20] Power, of course, is not evenly distributed between gaoler and prisoner, so that the prisoner's acts of assertion or aggression, in any playing out of cultural or religious conflict, must often mesh with his or her strategies for survival and take the form of resistance, which is not merely a passive act. To refuse food is both an aggressive and a reactive strategy. It is necessarily an act of resistance constructed within the terms set by the gaolers. Their legislation of eating is countered by Evans' own form of control of consumption. This serves further, then, to internalise the conflict, so that it is not only a battle of wills between subjects, but comes to be a battle between religious doctrines acted out within the subjectivity, bodily and psychic, of the prisoner.

The fact that it is food that is at issue, adds a further level of potency. To accept food is to accept the most psychically archaic, as well as culturally universal, of gifts. In eating the food of another (and all food, originally, in our personal psychic histories derives from our initial other – usually the mother) we incorporate that other. To refuse food is therefore to reject the contamination of the giver, by the refusal to absorb the other into the self. But this is not merely a matter of refusing a clearly defined other. If we turn to psychoanalytic accounts at this point, we find they offer differing interpretations of the significance of the acceptance and rejection of food.[21] Most useful for my purpose here is Julia Kristeva's theorisation of abjection – closely linked to its obverse, the ecstatic – as an aspect of primary narcissism. This details the significance of food in the constitution of a fundamental sense of self, but further relates this to the originating impulses of the monotheistic religious and cultural traditions of Judaeo-Christianity.

What Kristeva describes as the realm of the abject is situated precisely at that archaic psychic moment when the infant is at the brink of a series of differentiations and separations. It thus exists in a psychosomatic state where subject and object, inside and outside are as yet unmarked.

The experience of horror or disgust, the confusion or violation of boundaries between self and other or inside and outside are moments that later recall the abject:

> Abjection preserves what existed in the archaism of the pre-objectal relationship, in the immemorial violence with which a body becomes separated from another body in order to be.[22]

It is a psychic state, involving bodily reactions and manifestations, experienced at the level of the individual, but it is, too, a state to be negotiated culturally. Kristeva suggests that, 'Abjection accompanies all religious structurings and reappears to be worked out in a new guise, at the time of their collapse.' In particular:

> Abjection persists as exclusion or taboo (dietary or other) in mono-theistic religions, Judaism in particular ... It finally encounters, with Christian sin, a dialectic elaboration, as it becomes integrated in the Christian Word as a threatening otherness ...
> The various means of *purifying* the abject – the various catharses – make up the history of religions ... (p. 17)

Dietary taboos are a version of the prime separation narrated in biblical texts (the Old Testament) between God and humanity. Other crucial separations which differentiate the sacred from that which defiles it are versions of this initial separation, described in Genesis through the account of the eating of the apple in Eden and the Fall, which distinguishes the human and the divine. That which defiles is henceforth bound up with food, especially with flesh and blood. Kristeva traces through the Old Testament the progression from the prohibition against eating all flesh, to the dietary codes established in Leviticus. Through such organisations of pure/impure, Judaism marks its separation from the maternal cults which preceded it. In turn, Christianity transforms dietary regulation and related forms of taboo and exclusion into sin. The place of the sacred, however, remains the place of the undefiled, the pure, from which contamination must be expelled.

In Catholicism, then, Judaic dietary codes are reformulated temporally: food is regulated in terms of time, whereby meat cannot be eaten on Fridays, and fasting or limitation of food is confined to periods before Easter (Lent) and to the period before taking communion, thus purifying the body before the taking (and incorporation) of the host. Protestantism represents a further shift away from Judaism, in the replacement of notions of bodily purity by spiritual purity. In Quakerism, an extreme version of Protestant thought, where the Indwelling Light represents an immanent sacred, and where there is the possibility of the perfectibility of the believer guaranteed through the presence of the Inner Light, the pure/defiled negotiation regulated through distinction between types of intake (Judaism), differentiated periods of sanctity and human corruption (Catholicism), different people (the Elect and Reprobate in Calvinism) has no clear place or formulation. As in all antinomian religions, traverse of the realm of the abject is not regulated through doctrinal law, nor is the place of the sacred external. What seems to occur, then, is that such organisations or negotiations take place through individual psychosomatic manifestations. The abject – this fundamental series of simultaneous confusions and separations of the bodily and psychic, inner self and outer self, self and other, that are both individual and constitutive of the belief system of Quakerism (which is already informed by its oppositional nature, its need of a defining and opposed other to give it identity) – erupts, not as a retreat from self or identity, but precisely as identity. The Quaker subject does indeed become a 'living text', embodying and manifesting Quaker spiritual belief, which is located precisely at the site of the abject, or as Kristeva, speaking of the Christian mystic's 'familiarity with abjection [as] a fount of infinite jouissance', puts it, 'within a discourse where the subject is resorbed (is that grace?) into communication with the Other as with other' (p. 127).

The Quaker is an individual self, re-experiencing subjectivity at an archaic formative level and simultaneously embodying Quaker belief in its struggle to separate or exist. This is not a consequence of Quaker belief, but rather is its identifying and defining feature at this period. Quakerism is definitively situated at the brink of separation. Oppression or opposition are, in this sense, integral to its identity. Those separations marked out, in Kristeva's argument, between preceding religious formulations and a new religion (as in the case of the establishment of dietary codes in Judaism) occur in Quakerism as essential to its subjective experience. In Quakerism, one of the most individualist of all seventeenth-century sects, belief and doctrine are deeply enmeshed with the subjectivity of the individual so that doctrine does not exist as a set of legislations, but is manifest in the crises of subjectivity of Quakers. The opposing other is internalised and the battle of regulation and ejection is worked out across the body and soul of the practising Friend. Fasting, then, for Quakers in general, and for Katharine Evans in particular, has complex and potent meanings, and is central to the spiritual conflict that is the subject of *A Short Relation of Cruel Sufferings*.

The description of Evans' fast is interwoven, as I have remarked, with her and Cheevers' negotiations of the meanings of death, which arise directly from the expectation of Evans' death. The passage begins with Evans' rejection of a physician brought by the Friars to her in her weakened state:

> He said he had brought me a physician, in charity. I said the Lord was my physician and my saving-health. He said I should be whipped and quartered and burned that night in Malta, and my mate, too. Wherefore did we come to teach them? I told him I did not fear; the Lord was on our side. And he had no power but what he had received and if he did not use it to the same end the Lord gave it him, the Lord would judge him. And they were all smitten as dead men and went away.
>
> And as soon as they were gone, the Lord said unto me, the last enemy that shall be destroyed is death.
>
> And the life arose over death and I glorified God. The friar went to my friend and told her I called him 'worker of iniquity'.
>
> 'Did she?' said Sarah. Art thou without sin?'
>
> He said he was. 'Then she hath wronged thee.' (But I say the wise reader may judge.) (pp. 121–2)

Here we see a battle over the meaning of death that is directly linked to the political and religious contest between the women and the Friars, but which also extends the meaning of the fast and serves to amplify its significance as a means of establishing the function of sacrifice. At the beginning of the passage I have quoted there is a shift from threat of actual death to a colloquial simile ('they were all smitten as dead men'), which is then followed by a more general exploration of the symbolic, rather than the immediate, significances of death. There is then a multiplication of the meanings of death and life, beginning abruptly with the words from 1 Corinthians, which form part of Paul's explanation of the meaning of Christ's resurrection, here given as direct words from the Lord: 'the last enemy that shall be destroyed is death.'[23]

Evans' reference to these words then represents her sense of spiritual superiority and her political victory. By displacing the immediate events of her imprisonment onto a whole biblical history, she gives herself symbolic status as one who represents all those who are oppressed but are not subject to spiritual death. So, through the bodily act of fasting, the dispute over sinfulness, and through the much more textual exploration of meanings of death, Evans and Cheevers involve themselves in an evocation and dismantling of the central organisations of Christianity, most especially striving to understand and affirm the meanings of suffering and sacrifice (suggested by the references to Herod and the story from Genesis of God's replacement of Abraham's son Isaac by 'a ram in the bush') at the heart of Christianity:

> For between the eighth and ninth hour in the evening, he sent a drum to proclaim at the prison-gate. We knew not what it was, but the fire of the Lord consumed it. And about the fourth hour in the morning, they were coming with a drum and guns and the Lord said unto me, 'Arise out of thy grave clothes.'
>
> And we arose and they came to the gate to devour us in a moment. But the Lord lifted up his standard with his own spirit of might and made them retreat and they fled as dust before the wind. Praises and honour be given to our Lord forever. I went to bed again and the Lord said unto me, 'Herod will seek the young child's life to destroy it yet again.' And great was my affliction, so that my dear fellow and labourer in the work of God did look every hour when I should depart the body, for many days together. And we did look every hour when we should be brought to the stake, day and night for several weeks, and Isaac was freely offered up. But the Lord said he had provided a ram in the bush. Afterwards, the friar came again with his physician. I told him that I could not take anything, unless I was moved of the Lord. He said we must never come forth of that room while we lived and we might thank God and him it was no worse, for it was like to be worse. We said if we had died, we had died as innocent as ever did servants of the Lord. He said it was well we were innocent. They did also look still when I would die.
>
> The friar bid my friend take notice what torment I would be in at the hour of death: thousands of devils, he said, would fetch my soul to hell. She said she did not fear any such thing. (p. 122)

And finally the bodily exemplification and verbal assertion of meanings are split between the women, as Cheevers and the Friar argue about the meaning of Evans' apparently imminent death over her near-dying body. Cheevers articulates the meanings of Evans' bodily transactions for her. For a while the bodily and verbal coexist, although split between subjects, but, shortly after the end of this passage, Evans is moved by the Lord to eat once more:

The last day of my fast, I began to be a-hungry, but was afraid to eat, the enemy was so strong. But the Lord said unto me, 'If thine enemy hunger, feed him ... Be not overcome of evil, but overcome evil with good' [Romans 12. 20–1]. I did eat and was refreshed and glorified God. (p. 123)

The ambiguous or loose reference to 'thine enemy' (the Friars? Evans herself? something within Evans?) precisely articulates the complex negotiation of good and evil, self and other, that this passage as a whole demonstrates through its description of events and through its rhetorical shifts. The act of writing thus becomes a further means of patterning pure and impure. It is another form of action, linked to the verbal strategies of the women – their characteristic combativeness and their capacity to mingle divine and human voices into a narrative that is, at once, realist and ecstatic. It links, too, with their bodily enactment of the workings of the abject and the ecstatic. It is easy to make a link between food and words as oral activities which are implicated in all attempts to mark or renegotiate boundaries between self and other, inner and outer, bodily and spiritual. In addition to this, we witness in this passage a complex series of transactions occurring between ingestion, introjection of the Spirit and expression. *A Short Relation of Cruel Sufferings* fully exemplifies the implications of Quaker belief in the indwelling Spirit and its manifestation through the Word as living text.

*

Although each of the texts I have considered is produced out of an historically and politically distinct moment, they both describe a reaction to imprisonment that involves acute suffering and moments of ecstasy. Fasting in both comes to represent a condensation of strategies of response – a resistance, yes. But it also demonstrates a turning of an external conflict into a conflict of subjectivity, where good and bad, pure and impure, self and other, can be reconfigured through the return to archaic structures concerning separation within the individual. Imprisonment, fasting and ecstasy show us, in both texts, selfhood on the brink of simultaneous collapse and reformulation. The imprisoned subject takes wider, external conflicts into him or herself, although as the texts I have discussed show, for different reasons: Keenan seeks to find and possess meanings; Evans and Cheevers struggle to formulate and assert new meanings.

Imprisonment narratives and trial narratives, it can be argued, are deeply bound up with the history of Western autobiographical writing.

Carolyn Steedman has spoken fascinatingly of this very entwining of disciplinary narratives with the development of modern individualism, where the autobiography, linked to the novel, is the central and definitive literary form.[24] She speaks of the importance of the 'enforced narratives of the poor' at trials in the pre-history of the novel. But, I would argue, it is not only these enforced narratives which provide a significant inheritance. From the inception of modern self-writing in the seventeenth century we find also, along with trial records, the narratives of imprisoned subjects concerned with the possession of their own experience. These, like their twentieth-century inheritors, powerfully depict, at the level of content, processes of the disintegration and reformation of selfhood. At a wider cultural level we find that the specific conflicts described and the precise ways in which processes of disintegration and reformation are enacted point to specific crises in the history of individualism, and what it means at any historical moment to be an individual subject. For the writing subject we see, too, how the process of writing the self enacts a further level of taking apart and reconstituting the self in relation to reading others. Self/other relations are negotiated not only through the sufferings brought about by the conflicts described, but through the relation of self-writing subject and audience. For us as readers, then, participation in the suffering of a narrative may (alternatively or simultaneously) enhance our own sense of wholeness, as the safeguarding and complete other to whom the text is addressed, or engage us in a dialogue of suffering.

Foucauldian influences have encouraged us to see suffering in pre-individualist societies as being produced through ritual and spectacle and as serving to produce social bonding.[25] My suggestion is that in individualist societies the sharing of suffering is an equally forceful bonding mechanism, but one that is mobilised not by the spectacle and the sight of the suffering body as object, but by our, as readers, own negotiations of – empathy with, separation and distancing from – the suffering, disintegrating subject, returning to the most archaic structures of identity. Narratives of imprisonment figure as central to a tradition of autobiographical writings where, for writing subject and reader alike, the death, the birth and the autopsy of the self are simultaneously enacted, and where the individuality of any subject is revealed as inseparable from the historical and political context which vitally informs our individual and collective being at fundamental levels.

Notes

1 Brian Keenan, *An Evil Cradling* (London, 1992; 1993 edn) p. 1, quoting John 1:1.
2 *An Evil Cradling*, p. 1. All further references will be given in the text.
3 *An Evil Cradling*, p. 57. For an account of the significance of hunger-striking in Irish history, see David Beresford, *Ten Men Dead: the Story of the 1981 Irish Hunger Strike* (London, 1987) Ch. 1, especially pp. 14–15.
4 *An Evil Cradling*, pp. 220–1; p. 69.
5 I would assume that the text is understood differently in England and Northern Ireland, for instance.
6 Jean Baudrillard, *In the Shadow of Silent Majorities . . . Or, The End of the Social*, Semiotext[e] (New York, 1983) pp. 48–53, quoted in Maud Ellmann, *The Hunger Artists. Starving, Writing and Imprisonment* (London, 1993) p. 20.
7 Interestingly, Andrew Billen, in 'The Billen Interview', *The Observer Magazine*, 5 February 1995, reports that the actor Ralph Fiennes, in preparation for playing the part of Hamlet, has photocopied specific pages from *An Evil Cradling*. *Hamlet* has long been recognised as embodying *par excellence* those individualist values established in the Renaissance and inherited by following periods. It is perhaps not surprising that we should see a connection between an early modern drama of the isolated and alienated self and a late twentieth-century account of the self-in-crisis of a captive in Beirut.
8 John Bunyan, *Grace Abounding to the Chief of Sinners* (1666), ed. Roger Sharrock (Oxford, 1962) paragraphs 311–12.
9 Bunyan's *Grace Abounding to the Chief of Sinners*, perhaps the most famous seventeenth-century spiritual autobiography, was, of course, written in prison. In another typical confessional autobiography, *Satan his Methods and Malice Baffled* (1683), Hannah Allen, believing herself evil, identifies with criminals being taken to Newgate. See Allan Ingram's discussion of Allen earlier in this volume.
10 Barry Reay, 'Quakerism and Society' in *Radical Religion in the English Revolution*, ed. J.F. McGregor and B. Reay (Oxford, 1984, 1986 edn) p. 147. See this chapter and B. Reay, *The Quakers and the English Revolution* (London, 1985) for a full account of Quaker belief and the social and political circumstances of seventeenth-century Quakerism.
11 John 1:14.
12 (a) A.L. Morten, *The World of the Ranters* (London, 1970) pp. 18–19, speaks of the 'aggressive radicalism' of Quakers in the 1650s, although he suggests that this is lost in the 1660s. I would not agree that 'aggression' is uniformly diminished amongst all Quakers in the decade after the Restoration. (b) Christopher Hill, *The World Turned Upside Down: Radical Ideas during the English Revolution* (first published, 1972; Harmondsworth, 1975, edn) p. 239.
13 See *The World Turned Upside Down* and 'Quakerism and Society', for instance.
14 Katharine Evans and Sarah Cheevers, 'A Short Relation of Cruel Sufferings', in *Her Own Life: Autobiographical Writings by Seventeenth-Century Englishwomen*, ed. Elspeth Graham, Hilary Hinds, Elaine Hobby and Helen Wilcox (London and New York, 1989) pp. 121–3. Further references in the text are to this edition.
15 See 'Quakerism and Society', p. 148.

16 Diane Purkiss, 'Producing the Voice, Consuming the Body', in *Women, Writing, History 1640–1740*, ed. Isobel Grundy and Susan Wiseman (London, 1992) p. 45.

17 *Her Own Life*, p. 121.

18 Michel Foucault's enormously influential analysis of the meanings of forms of punishment in *Discipline and Punish: the Birth of the Prison*, first published as *Surveiller et Punir: Naissance de la Prison* (Paris, 1975; transl. by Alan Sheridan, Harmondsworth, 1979 edn) suggests a transition from spectacular punishments where the theatre of the body in pain prevails to reformative punishments, which begins in this period. The mission of the Maltese Inquisition is to reform and convert to Catholicism, as much as to chastise the prisoner through physical punishment and ultimately death. In experiences of imprisonment such as these, we see a mixture of spectacular and corporal punishments with a diversity of meanings. Foucault's argument that the body is inscribed in the discourses of the period and is experienced according to these has been equally – if not more – influential and underpins almost all the writing on the body which has proliferated in the past few years. Such ideas are assumed in this essay. See also *The History of Sexuality, Volume One: An Introduction*, first published as *La Volonté de Savoir* (Paris, 1976; transl. by Robert Hurley, Harmondsworth, New York, Victoria, Ontario, Auckland, 1978).

19 See *The Hunger Artists*, pp. 93ff.

20 Quoted from *Règlement pour la communauté des filles du Bon Pasteur*, in N. Delamare, 'Traité de police', 1705, in *Discipline and Punish*, p. 143.

21 We might turn, for instance, to Kleinian or Lacanian theories. For Kleinians the archaic world of the infant is envisaged through its relation to the good and bad breast. (See Melanie Klein, *Love, Reparation and Guilt, and Other Works 1921–1945* London, 1975.) To take food is ultimately to incorporate and physically to introject both good and bad aspects of the surroundings and caretaking environment represented by the m/other. At first the infant is not aware of itself as a distinct being, however. In order to become a 'self', with a sense of 'me' as distinct from 'other' or 'not me', it has to separate from its caretaking environment. Such processes of separation from the external other, necessary for the formation of an individual subjectivity, are bound up, thus, with the need to have taken that other into the self, to create a sense of self which *can* separate. In rejecting food, therefore, the subject may not merely be rejecting an already constituted other, but those very fundamental structures of self established through early processes of introjection and separation. To refuse food is to both purge the self, good and bad, and to explore the boundaries that constitute the self. For Lacan, picking up on Freud's interest in the circulation of currencies between subject and object (where milk is a primary currency between parent and infant) food is bound up with the subject's exchange values, through which he or she negotiates his or her boundaries in relation to others. (See *The Hunger Artists*, pp. 42–3.) The sense of control which Keenan, for instance, identifies in consequence of hunger-striking, emerges, according to such ideas, not just from identification with Irish history, or resistance of the other, or control of one's own body (although these are important) but from the re-establishment of boundaries between self and other.

22 Julia Kristeva, *Powers of Horror: An Essay on Abjection* (first published as *Pouvoirs de l'horreur*, Paris, 1980; transl. by Leon S. Roudiez, New York, 1982) p. 10. Further references are given in the main text.

23 1 Corinthians 15:26.

24 Carolyn Steedman, 'Other People's Stories: Modernity's Suffering Self', paper presented at 'Autobiographies: Strategies for Survival', The University of Warwick, October 1996.

25 *Discipline and Punish*, especially Ch. 2.

Index

Printed in the United States
By Bookmasters